Global Perspectives on Adult Education

Global Perspectives on Adult Education

Edited by Ali A. Abdi and Dip Kapoor

palgrave
macmillan

GLOBAL PERSPECTIVES ON ADULT EDUCATION

Softcover reprint of the hardcover 1st edition 2009 978-0-230-60795-8

First published in 2009 by PALGRAVE MACMILLAN® in the United States - a division of St. Martin's Press LLC, 175 Fifth Avenue, New York, NY 10010.

Where this book is distributed in the UK, Europe and the rest of the world, this is by Palgrave Macmillan, a division of Macmillan Publishers Limited, registered in England, company number 785998, of Houndmills, Basingstoke, Hampshire RG21 6XS.

Palgrave Macmillan is the global academic imprint of the above companies and has companies and representatives throughout the world.

Library of Congress Cataloging-in-Publication Data

Global perspectives on adult education / edited by Ali A. Abdi and Dip Kapoor.
p. cm.
Includes bibliographical references.

1. Adult education. 2. Education and globalization. 3. Adult education—Case studies. I. Abdi, Ali A., 1955– II. Kapoor, Dip.
LC5215.G57 2008
374—dc22

2008021620

First Edition: January 2009

ISBN 978-1-349-37472-4 ISBN 978-0-230-61797-1 (eBook)
DOI 10.1057/9780230617971

Design by Westchester Book Group.

10 9 8 7 6 5 4 3 2 1

Transferred to Digital Printing 2012

Contents

Preface

This book is a breath of fresh air for Western critical adult educators who by definition are trapped in their own contexts no matter how they struggle for the critical edge. It is exciting to see a book like this make it into print in English. For the most part, the contributing scholars—except for one—have been born and nourished in the "global South," thus having direct social and historic roots with the experiences they write about. It is true that most are Western educated and some have spent years outside the culture they write about; however, for me, a seasoned Western popular educator, it was an adventure to immerse myself in this book and to appreciate the insights and perspectives provided by these authors who represent their own cultures and speak from them.

The volume is composed of two parts: one that features concepts and a second that presents nine case studies. Critical analysis appears in most chapters. In the first part, the definition of *adult education* is firmly established as including social goals as well as instrumental ones. The pressure of international financial institutions to narrow educational goals to ones of material production is documented in many settings. While this disparity is laid out conceptually in the first six chapters, the case studies reiterate the problem. For example, poor Buddhists in Bangladesh have a 180-degree value difference with Western development. They view dam building and the green revolution as a ravaging attack on their moral system, resulting in the loss of indigenous knowledge and the destruction of the social fabric of villages. This case study is presented so starkly that the reader can't help but see the absurdity of value imposition in the name of development on the core values that form identity.

The importance of social development and citizenship in a global setting as the goal of education is the standpoint of this book. As Shizha and Abdi state, "Due to globalization, adult education changed from a community-based cultural and social enterprise to an individualistic and egoistic project. Altruistic adult learning has been overtaken by individual indulgence at the expense of community and national development."

What are their alternatives to the Eurocentric individualistic and egoistic adult education project? I will discuss five: subaltern social movements; political education as citizenship; training programs for community activists; critiquing our methodology; and understanding popular uprisings.

Subaltern Social Movements

The reconceptualization of social movements as an important advance in social movement theory is discussed. From a southern perspective, subaltern social movements (SSM) cannot be subsumed in the categories of new social movements (NSM) or global social movements (GSM), nor lost in the dichotomous old social movements (OSM) versus NSM. Six distinctions are outlined for SSMs for our analysis. With this framework, we now have a way to examine learning within SSMs, an important contribution to social movement learning theory.

Political Education as Citizenship

Rarely are adult educators from the South studied, with the exception of Paulo Freire. One of the strengths of this book is the introduction to a number of adult educators of national stature. Two prominent educators are foregrounded for us: Julius Nyerere and Freire. Each is accorded a biographical chapter in Part 1. Freire is well known and several chapters draw on his work. In a chapter demonstrating how Nyerere is relevant to African education today, the authors discuss Nyerere's cultural approach to adult education with a program called *Ujamaa*, which promoted participatory democracy and self-reliance in Tanzania. Several of the contributors see the potential of utilizing indigenous ways of knowing and a revival of African pedagogy such as that developed by Nyerere for Africans to extricate themselves from colonialism and global intrusion.

Other national leaders discussed include Confucius and Mao Zedong, although this is more of a discussion of John Dewey and Confucius, with a nod at Mao as an adult educator in the formation of the present Chinese educational system. However, Mao was committed to the education of the peasant and developed worker's colleges, challenging the elitism of Confucian-based education. Included among the lesser known educators are Eric Williams, who used the public square as a means of giving the people political and civic education (he called it the

University of the Public Square and it was through his political speeches that he built a critical understanding among the People's National Movement in Trinidad/Tobago); Walter Rodney, the Guyanese scholar and noted author of *How Europe Underdeveloped Africa* and *Groundings of My Brother*; and, finally, Maurice Bishop, the Grenadan leader of the New Jewel Movement who saw political literacy and the ability to read and write for all citizens as basic to freedom. These political leaders are held up as exemplars for educating the citizenship. With Mao and Nyerere, in particular, we have examples of two individuals attempting to establish a national socialist system of education, and in whose ideas some African and Chinese scholars see potential for today's problems.

Training Programs for Community Activists

There is one case study from Indonesia where a training program for popular educators, based on Freiean principles, has been established. Noting the many types of nongovernmental organizations (NGOs), local activists got together and, through critical reflection, established a more transformative paradigm for the indigenous people. This was implemented through a school to retrain community social activists in critical pedagogy. The methods and curricula are provided in this chapter, as well as descriptions of participants. It is noteworthy that it started with a critique of one's own practice; many NGOs do not realize that they have become complicit with the power structures.

Critiquing Our Methodology

From Brazil we have a report of the street children and a strong request for more ethnography in our Freirean approach. In reading this chapter, I couldn't help thinking of the prize-winning television series, "The Wire," that tells the story of youth gangs in Baltimore. After five years in the series, no one wins; what you realize is that it is almost impossible to escape the social structures that define us. It is the same way the children understand the culture of the Brazilian streets and why they couldn't care less about traditional literacy or mainstream values—only survival. It was only through ethnography that researchers discovered whether agencies that were supposed to be helping were being useful or being used. Again, like the activists in Indonesia, the Brazilian street workers discovered the deficiencies in their own and other NGOs through critical examination. They did this through ethnography. Only by understanding these relationships could the situation change so that the street children could become knowledgeable, and cultural producers in their own right.

Understanding Popular Uprisings

The two Argentinean case studies included here are fascinating. One describes the popular uprising of hundreds of thousands of Argentineans who organized themselves into informal groupings as a result of the economic crisis that occurred in 2000. It was "a quick formation of a social movement" (chap. 13) that, even after a return to normalcy, remained independent of formal structures. This macro description of the spontaneous mobilization of an entire modern society is important in and by itself. Why it continues and why it stays apart from public structures today when the emergency is past are important questions. It is a phenomenon that bears watching as their educational programs maintain a critical edge.

The second study is about the Madres Movement. Most of us are now familiar with that movement but we are provided here with a Freirean analysis of how it developed and how conscientization among members occurred. This analysis of a well-documented social movement provides us with details of the origins of the movement among the women and, subsequently, how politicization occurred. It is fascinating because there is no outside organizer. Like the larger mass movement, the initiation is from the people themselves.

Conclusion

I find this book to be an important contribution to the literature of adult education and a unique contribution to critical pedagogy. North American scholars would be well advised to make this book required reading in their graduate programs, regardless of the study focus. In fact, Human Resource Development programs especially should have their students become aware of the relationships among resource extraction, development, and exploitation (see Chapter 11 on gold mining in Ghana). All adult educators should debate the inclusion of social goals in the modern definition of adult education from the perspective of the South, as well as the North, and give consideration to putting a broader definition into practice.

Adult educators in the South can find much to celebrate in the contents of these chapters. One would hope others will be inspired to continue to bring more information forward, to research more of their histories, to study more of their social movements, and to learn and share this knowledge and these experiences with others. I sincerely hope there will follow many more books such as this one.

Dr. Phyllis Cunningham
Northern Illinois University

Acknowledgements

Putting together an edited book involves the collaboration of so many people, and this book was no exception. What prompted us to think about creating a reader on international adult education was the observation that there were conspicuous gaps in the current literature on adult education in terms of perspectives, analysis, and authorship from the countries of the global South (or the "developing world") and diaspora from these regions. We sought to begin to address this oversight in the hope that this will encourage others to continue to shape this path of analysis and exploration, with the view to enrich the discussion around adult education in Southern contexts and its wider implications for adult learning, social development, and change in global contexts, because learning and education in their varied cultural and historico-political locations can no longer be seen to be the sole prerogative of adult education experts from Northern contexts alone. More importantly, this collection would attest to and foreground the continued significance of internationally acclaimed approaches to adult education with Southern roots (theoretical and practical), an adult education that speaks to processes of global-local (dis)enfranchisement of adults in the African, Asian, and Latin American/Caribbean regions. Informed by such lines of thinking, we began the search for prospective contributors or people whom we felt would strive to strike a balance between critique and possibility and despair and hope, while being cognizant of the ever-present challenges posed by global social relations of power—relations that have been in the making over the past five centuries—and their ongoing implications for those being continually relegated to the margins.

In retrospect, we can confidently state that all contributors to this collection have been more than up to this task and it is through their committed support and scholarship that the book is finally here. We would like to thank each of the authors for their effort and their contribution toward this timely analyses of adult education and social development in the South—indeed, a set of contributions with pertinent implications for global contexts. We would also like to thank our editorial assistant from the Department of Educational Policy Studies here at the University of Alberta, Lois Edge. Lois's excellent and meticulous editing skills were exceptionally helpful in producing a work that meets, as much as possible, our expectations and, hopefully, the expectations of those who will read this book. Thanks to Julia Cohen of Palgrave for her patience and enthusiasm for this work and for her continuing support in helping us meet the requirements of the publisher. We are also deeply grateful to Professor Phyllis Cunningham, who kindly agreed to pen the preface despite the pressures on her time. We thank the anonymous reviewers who have evaluated this work for the publisher. Their comments and insights were helpful in ameliorating both the descriptive and analytical platforms of the book. As a result of their input, we can say that this is a better and more inclusive collection. Last but not the least, we are grateful for the continuing support of our families. Without their patience and understanding, this project would not have come to fruition, and we are always cognizant of how central they are to all our achievements.

Notes on Contributors

Ali A. Abdi is professor of international development education in the Department of Educational Policy Studies at the University of Alberta in Edmonton, Canada. He is president of the Comparative and International Education Society of Canada (CIESC) and has coedited (with Korbla Puplampu & George Dei) *African Education and Globalization: Critical Perspectives.*

Charlotte Baltodano is a human rights lawyer from Nicaragua. She did her master's in comparative law at McGill University. Currently she is doing her doctoral studies in international development education at McGill University. Her present research focuses on rural women of South India and issues of empowerment, agency, and change.

Bijoy P. Barua (Ph.D.) is currently associate professor and chair in the Department of Social Sciences at East-West University, Dhaka, Bangladesh. He offers courses at both graduate and undergraduate levels in the areas of development studies, research methodology, civil society, gender and development, and sociology. He has also worked as researcher, program manager, and consultant in Bangladesh, India, Nepal, Sri Lanka, Thailand, Vietnam, Ghana, Switzerland, and Canada. His research interests include international development, comparative education, participatory research, ethnic minorities, indigenous knowledge, engaged Buddhism/ecology, community health, gender and culture, development and civil society. He has published in academic journals such as *International Education* (USA), *Canadian Journal of Development Studies, Medicus Mundi* (Switzerland), *Managerie* (Germany), and *Parkhurst Exchange* (Canadian Medical Educational Journal for Doctors).

Luis-Alberto D'Elia is a former prisoner of conscience from the Argentina of the 1970s. A clinical biochemist, both he and his wife were imprisoned for their social justice work. He is currently a lecturer in education at the University of Alberta, and is also the police and human rights coordinator for Amnesty International, Canada.

Marcelo Diversi is a faculty member in the Department of Human Development at Washington State University, Vancouver. His work has centered on the intersection between identity formation and educational aspirations of disenfranchised youth. He has conducted ethnographic research with Brazilian street youth and Latino immigrants in the United States.

Shibao Guo is assistant professor in the Workplace and Adult Learning Program (WAL) within the Faculty of Education at the University of Calgary in Canada. His research interests include social justice and equity in education, multicultural and antiracist education, adult education and community development, citizenship and immigration, Chinese immigrants in Canada, and comparative and international education.

Dip Kapoor (Ph.D.) is associate professor, international education, in the Department of Educational Policy Studies, University of Alberta, Canada, and president of a Canadian voluntary development NGO working with scheduled tribes and scheduled castes in eastern India. He is a volunteer research associate at the Center for Research and Development Solidarity (CRDS), Orissa, India. His scholarly and applied interests include sociology of education & development; globalization, social movements, and popular education; global education; and qualitative/participatory research. His articles have appeared in *Convergence; Adult Education & Development; Journal of Postcolonial Education and Canadian & International Education.*

Michael Tonderai Kariwo is a Ph.D. candidate in the Department of Educational Policy Studies, University of Alberta. He was registrar of the National University of Science and Technology in Zimbabwe from 1991 to 2003, and director of planning at the University of Zimbabwe from 1989 to 1990. He is also a former high school teacher and principal.

Valerie Kwaipun is a Canadian activist/writer. Her current research focuses on the role of learning in subaltern struggles with development-led displacement in Africa. She recently obtained a Master of Arts degree from the Department of Integrated Studies in Education at McGill University in Montreal. Valerie resides in Accra, Ghana, and continues to work closely with the Wassa Association of Communities Affected by Mining (WACAM).

Peter Mayo is professor in the Department of Education Studies at the University of Malta. His areas of research are adult and continuing education, and sociology of education. One of the foremost scholars on Freirean and Gramscian philosophies of education in the world, his numerous books and articles can be viewed at http://www.educ.um.edu.mt/mayo.

Christine Mhina received her Ph.D. in international/intercultural education from the University of Alberta. She is now a university instructor for Swahili language and culture in the Department of Modern Languages and Cultural Studies. Her research interests include participatory democracy, women's emancipation, community-based action research, and grassroots initiatives.

Muhammad Agus Nuryatno (Ph.D.) received his doctorate in education from the Department of Integrated Studies in Education (DISE), McGill University, Canada, and is currently a faculty member at the State Islamic University Sunan Kalijaga Yogyakarta in Indonesia. He has published articles in numerous Indonesian educational and social development journals and in the journal *Convergence* (International Council for Adult Education). His research interests include critical/popular education, Islamic education, and critical pedagogy and cultural politics in the Indonesian context.

Ladislaus Semali is an associate professor of education at Pennsylvania State University. His research interests include the comparative study/analysis of media languages, contexts of cross-cultural literacy curricula, and critical media literacy across the curriculum. He is the director of the International Consortium on Indigenous Languages (ICIK) and has coedited (with Joe Kincheloe) *What Is Indigenous Knowledge? Voices from the Academy.*

Edward Shizha received his Ph.D. from the University of Alberta and is currently assistant professor of education at Wilfred Laurier University in Ontario, Canada. A former lecturer in sociology of education at the University of Zimbabwe, his areas of research include sociology of the curriculum, globalization of education, indigenous knowledge and formal schooling, and education and social justice/equity issues.

Samuel Veissière is assistant professor of anthropology and social sciences at the University College of the North in Thompson, Manitoba. He has conducted ethnographic fieldwork on subaltern agency, street livelihoods, and transnational desire and violence in Salvador da Bahia, Brazil. His research interests are centered around questions of postcolonial and transnational domination and resistance, and learning from criminalized forms of spontaneous subaltern conscientization, mobilization, and/or rebellion. He is currently working on a book that combines his Brazilian fieldwork with a follow-up investigation of Brazilian sex workers' livelihoods in Europe, and is involved with community-based First Nations education in Canada/Turtle Island on an ongoing basis.

Jean Walrond received her elementary and secondary schooling in Trinidad and Tobago in the West Indies. Her postsecondary education in Canada includes a Ph.D. degree in education from the University of Alberta. She teaches in the Department of Sociology at Concordia University College of Alberta in Edmonton. Her research interests include cultural studies, sociology of education, multiculturalism, antiracism education, and gender and representation.

CHAPTER 1

Global Perspectives on Adult Education: An Introduction

Ali A. Abdi and Dip Kapoor

It is generally assumed that adult education involves select reconstructions of one's life needs and possibilities relative to the world in which one resides at a given time and with respect to specific changes in the sociopolitical, economic, and technological lives of societies. Also located in the realm of adult education is the andragogical assumption regarding the adult learner's greater level of voluntarism and autonomy in the learning undertaking. As pointed out by Julius Nyerere (1968), whose brand of postcolonial adult education for self-reliance in Tanzania was unique in many ways, every adult education learner knows something, not only about the demands of the context in which the desire for more learning is created, but also about the subject that she or he is interested in, even when the learner may not be aware of that *a priori* knowledge. In terms of global perspectives on adult education (the focus of this collection), global discussions, analysis, and approaches should be informed by the situational characteristics of each locale and by extension, should pay attention to what can be shared with other adult education programs and nonprogrammatic adult learning throughout the world. Additionally, teaching and research in global and international adult education always contains an element of comparison, as our understanding of each new initiative persuades us to see it in relation to what exists around its instructional borders.

In the way we are deploying it here, adult education, in terms of its contents, objectives, and purposes, fits the widely used definitions that address teaching and learning realities of people who want to improve their life situations beyond the context of what is generally termed initial education. The insertion of social development here is deliberate in that all types of education should, for us, at least instigate some possibilities for people's well-being in the different relationships and locations of their interactions. This might suggest a need to utilize adult education programs to economically advance people and socioethnic groups traditionally excluded from processes of learning and credentialing; for example, programs to assist with career advancement and to provide better opportunities. One may go further to suggest that adult education for social development also involves revolting against aspects of the local or universal status quo and thereby higlights the importance of adult education for political development and consciousness-raising, especially among those who have been marginalized by the dominant world system.

In terms of adult education for politicization, two of the most important writers in the field, Paulo Freire (1985, 2000/1970, 2001) and Julius Nyerere (1968, 1974), have emphasized adult education as the terrain on which to create citizens who are aware of their oppression or underdevelopment—an adult learning process that can precipitate and sustain their mental and material liberations. These contributions are even more vital at a time when we have to deal with the realities of globalization and all this entails in the manner of citizen's rights to learn and live productive, examined lives.

Globalization and Global Contexts of Adult Education

The ubiquitous descriptive and analytical insertions of the now explosively used term *globalization* and its operationalization in most works that deal with educational and social development issues is anything but accidental. Recalling a discussion one of the authors had on the topic of why so many graduate students in African universities were writing their theses on globalization, an African student visiting Canada in the mid-1990s responded that globalization has a direct effect on the daily lives of people, and as long as that is the case, researchers will continue to talk about it, critique it, and, where needed, analytically deconstruct it even if its economic categories are too strong to be disturbed by academics and their occasionally inspired students. As such, intentionally included in this volume (see, especially, Shizha & Abdi's and Kapoor's chapters) are descriptively interlinked ways of rethinking and analytically relocating current forms and contexts of international adult education that could effectively respond to growing educational and social development pressures brought about by the current world system. As Wallerstein (2004) noted, the interconnectedness of this world system is deeper, more direct, and has more impact on people's lives than many of us might realize prima facie. Indeed, the complexity and historical and contemporary interconnectedness of the institutional, organizational, disciplinary, national and extranational, and even familial structures of the world system are more expansive and intricate than many of us might realize.

Interestingly, the changes brought about by globalization, even if they have hastened the wide-scale exchange of goods and services, have not addressed the needs of those around the globe whose prospects, in terms of general life viability and social development, have been diminished by the openness that was supposed to benefit all (Abdi, 2006; Rahnema, 1997; Stiglitz, 2002. As Harris (1996) notes, adult education programs have so far tended toward finding applications for new global technologies rather than creating spaces of learning that could effectively respond to social needs and expectations. Further, as Pöggeler (2002, p. 107) says, "Adult education has [recently] begun to investigate the question of how it is globalizing itself and how much it is involved in the globalization of individual and social life." With this in mind, the main issue should not be the globalization of adult education per se, but to what extent that expansion constructively contributes to the lives of people. If some perceptions of globalization, as Pöggeler notes, were about the exchange of human values and understandings, and possible inclusive collaborations where all have a discernible and tangible stake, then we propose that there should also be multiplicities of learning characteristics and possibilities that color and inform that process. In its modernity-driven rationalization, however, globalization seems to have led instead to social division and cultural fragmentation (van Gent, 2002).

Despite such disappointments, globalization is an expansive reality. As such, Johnston (2003, p. 4) should still be right when he writes that "the rapid pace of global economic change brings with itself a seemingly irrefutable need for lifelong learning," with important implications for the need to design it in a way that develops human capital. Questions would still remain: how can this be done, what are the final intentions, and who benefits? With people around the world having different needs, and different understandings of adult education, one should be careful in suggesting specific globally located possibilities. In that regard, we agree with Fenwick and Tennant (2004, p. 55) when they write, "There is no generic, essentialized adult learner who can be described in ways that accurately and responsibly portray the myriad differences between people and the changes they experience. Indeed, ideas of adulthood vary so widely that announcing adult learning as a unique and distinct category has become a dubious enterprise." As such the usually decontextualized dominant streams of the globalizing process and its adult education categories (as well as other types of education) have not enlarged the pie for inclusive social development. Hence, our possible return to the earlier discussion on the need to rethink the case and to promote effective international adult education that will enhance global human well-being.

Outside the educational arena there have been some positive outcomes of globalization, including the use of new information technologies and other platforms of globalization to expose human rights abuses as well as the ongoing oppression of (not necessarily numerical) minorities. In terms of primary economic categories, we should not be fooled by the few so-called narratives of the success of globalization; in fact, even in places where conditions might have improved for some, the situation has usually gotten worse for many others in the same context. Consider, for example, China, where the "economic miracle" of globalization has relegated 300 million of its citizens to extensive poverty and environmental deprivation while a significantly lesser number of the citizenry have benefitted considerably. Perhaps we need to rethink the many assumptions we are making about that situation. Look too at the emerging economic power of India. One can hastily point to the willingness for open commerce and the harnessing of human labor to uplifting the people, but the industrialization of India is being achieved at the expense of rural people and poor peasants whose lands are being expropriated for the benefit of a myriad of multinationals (see Kapoor, this volume) who are establishing their manufacturing and distribution outlets there and in other places where the exploitation of cheap labor and other resources are the perennial common denominator. If that is the case for China and India, then the reality for others in the grossly misnamed "global village," even with all the constructive intentions we could display, is bleaker than ever.

In the African context, for example, the full-blown external and internal effects of globalization, roughly covering the period from the early 1980s to the present (see Held, 1999; Tikly, 2001), has led to an overall decline in economic, educational, and other social development conditions (Abdi, 2006). And Africa, where civil strife and other systemic calamities may have derailed any possibility of viable forward movement, is not alone in its response to globalization. In fact, among countries that have jumped on the globalization train, the majority were worse off in the year 2000 than they were in 1990 (UNDP, 2003). Now, with the current global economic, energy, and environmental crises, which are mostly driven by the mantra and practices of globalization, and the open and unencumbered competition for everything on earth and in outer space, the prognosis is anything but encouraging. As such, it is our understanding that a proactive understanding of the global contexts of adult education will help ascertain the crucial intersections of educational and social well-being that are greatly needed, especially by those whose small fishing boats are being sunk by the ocean liners of globalization.

Beyond its ongoing descriptive and analytical threads, this volume aims to represent the perspectives of those who write from Southern contexts (not in the geographical sense, but in their scholarly intentions), and whose understanding of these locations are either historically tenable and/or culturally attached. The purpose here is not to deploy a Cesairean point (Cesaire, 2000/1947) of the *pays natal* genre, rather, it is to speak about perceived commitments to constructively interact with learning realities and needs of the situations under consideration. It should be understood, that adult programs in most of the developing world (global South), the areas this book focuses on, are designed, like other programs of learning; they are based on the histories and philosophies of colonial educational projects that neither advanced nor appreciated the cultural or linguistic locations of the communities they presumably served. While the linguistic point is more complex than the cultural one here (i.e., more difficult to deal with because of the lack of readily available resources and the heavy entrenchment of European languages in all forms of education), the analytical dimensions of this book should, minimally, represent a counterdirectional response that conveys new messages for a better understanding of the situation. The primary reason the colonial side of the story is continuously important for us (even if it becomes selectively redundant for some otherwise intelligent people) is that the earliest colonial educational programs trained indigeneous adults in languages, religious proselytizing, and other skills to support the effectiveness of the colonizing power. It is therefore important to place any international adult education analysis on the historical platform of colonialism.

In the continuing attempts to recast the philosophical and cultural foundations of adult education, one must consider the relevance of Julius Nyerere (1968, 1974). Nyerere, Tanzania's first postcolonial president and well known as a philosopher-statesman (see Mhina and Abdi, this volume), critically understood what colonialism has done to the educational and social development situation of his country and to the wider African world. To Nyerere, it was not only the destruction of these systems that derailed the collective life systems in subsequent *vitae Africains*, but the philosophical and psychocultural distortions that were achieved by colonial domination. Extensive analysis of these important issues of learning and social lives by others who have studied these contexts support Nyerere's ruminations (see, among others, Abdi, 2002; Achebe, 2000; Rodney, 1982; wa Thiongo, 1986). When one considers colonialism as being first and foremost a psychological project, supported by cultural, educational, political, economic, and technological dimensions, deconstructing people's perceptions of their world, regardless of the physical relationships, becomes important.

Colonialism more or less has had the same effect on educational and overall life relocations in other places where it has dominated. In the Asian context, for example, Macaulay's oft-referenced "A Minute on Indian Education" (1995/1935) speaks volumes about the mental depatterning that was important for the long-term educational, cultural, and linguistic domination of the European metropolis. As we said above, the process always seems to start by the diminishing of the epistemic and epistemological worlds of the to-be-colonized natives, which become the sine qua non for success of the colonizing project. In Macaulay's analysis, once the "uselessness" of Indian languages was promoted, the rest was easy. Macaulay's recommendations were that the British needed to create "a class who may be interpreters between us and millions whom we govern; a class of persons Indian in blood and color, but English in taste, in opinions, in morals, and in intellect. To that class we may leave it to refine the vernacular dialects of the country, to enrich those dialects with terms of science borrowed from the Western nomenclature, and to render them by degrees fit vehicles for conveying knowledge to the great mass of the population" (p. 430). What Macaulay was talking about, especially in the latter sections of that quotation, is not unrelated to what Memmi, in his classic *The Colonizer and the Colonized* (1991/1957), calls the perforce established collaboration between the two figures of the colonial story. The story, as we should know by now, contains, mainly via its educational intersections, the deliberate mental colonization of its victims (Nandy, 1997; wa Thiongo, 1986), which any counterhegemonic learning possibility should critique and neutralize. Moreover, the pointers on globalization and its expansive practices should not be detached from colonialism. In fact, many would see colonial education as the first highly organized form of globalization, with most of the worldwide dominant and subordinate relationships, including today's economic placements among individuals, groups, and nations, selectively emanating from that. As such, any viable discussions that deal with the educational problems, developments, and prospects in previously colonized spaces of our world have to engage new discursive formations about the historicities and multilocational actualizations of what shaped the contexts we are describing. It was indeed in response to these historical and contemporary realities that Africa's most important adult educator, Nyerere (1968), suggested one goal for the indigenized learning projects: "Our education has to foster the social goals of living together and working for the common good. . . . [It] has to prepare people to play a dynamic and constructive role in the development of a society . . . [and] it must inculcate a sense of commitment to the total community." Although the points here are both philosophical and cultural, it is important to note that there are also some important citizenship considerations. Should international adult education programs promote individual rights vis-à-vis collective rights? Should the objectives of social development be focused on uplifting the lot of all in the society, or should there be more talk of creating open spaces of competition that are so ingrained in the politics of Western traditions?

The answers to these questions are not easily achieved. They should not be rhetorical or ideological, nor should they be just historically located or culturally exclusionist. Despite the legitimate and important disquisitions we have related with respect to colonialism and globalization, the practical point is that we now live in an environment where the world is more intermeshed than ever. In this complex, interactive, highly tentatively *mondo moderno,* even those whose pasts and presents we are trying to be sympathetic toward would tell us that they are not looking for the primordial order, at least not at this time. We need not try to box them (even if we were able to do that, which we cannot) into a preferred analytical straightjacket that would imprison them in historical spaces where they cannot re-create themselves and their world. As Fanon (1968) so brilliantly puts it, while one can understand their past as effectively as possible, dwelling on it too much might not be the best choice for people's actual well-being. Again, the historical analyses are important, for they define the rescinding of people's citizenship rights without which educational progress may not be as meaningful as it should be. This really highlights the need for adult learning programs that are contextually responsive. Even Eduard Lindeman (1961), one of the earlier writers in the field, clearly saw the enhanced agency of adult education as an important platform for social progress provided that it effectively responded to the realities of a continuously changing world.

Adult Education as a Forum for Citizenship Development

The role of citizenship and its practical implications and possibilities are especially important to us. Indeed, the incredibly simple equation we are developing here, that is, education leads to social development (E = SD)—no mathematical formulae, just a symbolic depiction—has at its roots the re-establishment of those basic civic rights that have been usurped by others, including colonial powers, postcolonial elites, and the apostles of the currently expansive neoliberal paradigm that is being pushed by multinational corporations and their Western governments. As such, the discussions contained in this book directly or indirectly push for viable citizenship possibilities that liberate people from the false rhetoric of the domination and control, for without that happening, our own intentions of analyzing education as a primary agency of human well-being will not be effective. We believe that it is through informed critical citizenship that people can achieve better livelihood realities. For example, the physically located violations of citizens' rights, such as illiteracy and poverty, are first and foremost denials of citizenship rights that should be resolved along with all other rights that should be enjoyed by all. John Dewey's (1926) observation that maintaining effective political systems (democracy, in his understanding) needs not just any education, but a type of learning that recognizes and manages the interpersonal, intergroup, and international clashes that would be intrinsic to the lives of human beings, is important in this context. Thus, education directly affects not only the nominal locations of citizenship, but also the all-too-important attainment of active citizenship (as opposed to passive citizenship in the Kantian sense), which requires select knowledge and skills that can bestow the right political cultural capital on citizens (see Saha, 2006).

Significant to the "correct" globalization of citizenship understanding, especially in its intersections with democracy, are the rights of the individual. Furthermore, one of the most vital aspects of Western liberal democracies is governmental mediation of public spaces so that people can compete with little interference from state apparatuses. In any citizenship discussion of international adult education, as Nyerere says above, such an understanding of citizenship and democracy is not common to the majority of the world's population. Where the masses, for example, are oppressed by small postcolonial elites, especially when that oppression has been endured for decades and sometimes for centuries, a concerted effort to free the minds as well as the bodies of the people will be paramount. This step-by-step recitizenization and, where viable, democratization of public life for all is reflected in the work of such mass liberation theorists as Paulo Freire.

Freire discussed and analyzed the problems as well as the possibilities of people selectively looking inward to critically understand the histories and actual reasons for their oppression. Freire's body of work (the complete or partial focus of chapters by Mayo, Nuryatno, Kwaipun, and Baltodano in this volume) is extensive and widely read. Although we may safely assume that many readers are familiar with it, a brief look at his magnum opus, *Pedagogy of the Oppressed* (2000/1970), reveals that Freire popularized his important process-oriented concept of "conscientization" to trigger action and reflection upon the world to transform and, in the process, achieve the right praxis for community emancipation and development. For us, especially, one important bridge in Freire's analysis of the context is his critical understanding that to achieve effective citizenship through popular education, the necessary temporal and learning efforts must be harnessed and expended—which affirms the maintenance of the goals achieved and the education as well as the conscientizing processes that inform it. To do so involves citizens ascertaining an inclusive historical consciousness, educating themselves (see D'Elia, this volume), analyzing problems and opportunities as they emerge, and eventually acquiring the pragmatic confidence to solve their citizenship woes, not necessarily by themselves only, but by enlisting the support of their former oppressors as well. As popular education sites are usually the loci of the Freirean pedagogies, the need to learn in order to act, and to act in order to learn (see Crowther, Galloway, & Martin, 2005), affirms the actualization of one's humanity, which should be built on the word-work-action-reflection continuum that can create and sustain the needed pedagogies of liberation, conscientization, and overall human progress. In the international education efforts presented in this book, it is hard to separate the discursive trajectories and analytical intersections from the desirable citizenship platforms that are either missing in the lives of people or are incomplete in their social development structures and outcomes.

In our observations, we are deliberately extending, as we indicated above, the boundaries of citizenship in the sense that all rights or responsibilities and the myriad attachments that form and inform all qualities and dispensations are important items of life, which all people in the world can lay claim to. The focus should be about what we might call radical platforms of citizenship, such as selectively indigenizable democracies that are not, at least theoretically, limited. Thus, the relevance as well as the importance of local African knowledge systems for the educational well-being of people (see Semali, this volume) can actually be seen as a fundamental citizenship right, for it recultures and sustains the primary needs of life as opposed to potentially alienating people and adding to deprivation of learning and progress. Interestingly, rights that some might rank as even more important (access to land/food, clean drinking water, shelter, and security) are being violated all over the world, but the interconnectedness of *all* rights should not be violated, for in simple terms, once some are violated, the rest could follow. Supporting people's rightful claim to indigenous knowledge and ways of knowing may deflect some of the historical realities attached to colonial-induced, long-term disturbances and destructions (i.e., direct and primary violation of basic human rights) of people's ontologies, ecologies, and overall life management systems.

To go back to the previous point, the establishment of radical rights platforms (intended here to be as expansively inclusive as possible) through international adult education programs or via other projects of learning is important and is reflected in the contributions to this work. Indeed, the acquisition of all knowledge, as Dewey (in Jarvis, 2004) notes, should have humanistic intentions, which would then add value to its relevance for enhancing people's capacities to function as productive citizens of their societies. This last point should not be detached from the role of education in establishing positive citizenship rights that mediate and guide the way we understand, define, and practice civic duties and relationships. As Mouffe (in Coare & Johnston, 2003) notes, the type of citizenship rights and platforms we create will have a lot to do with the type of society and political community we desire to create and live in. With all education having important elements of civic training, the desirable political community will be densely reflective of both the adult and initial learning contexts we experience for the ever incomplete citizenization of the population.

Organization of the Book

Following this introductory chapter are fourteen chapters that cover topics ranging from general analysis of globalization and indigenous knowledge systems to specific country/area case studies drawn from Africa, Asia, Latin America, and the Caribbean. In Chapter 2, "Globalization and Adult Education in the South," Edward Shizha and Ali A. Abdi examine some of the intersections of globalization and adult education in relation to historical and emergent trends, with particular reference to African contexts and perspectives. Shizha and Abdi discuss the social role of adult education with respect to civil society initiatives, which aim to achieve democratic reforms meant to enfranchise the world of marginalized groups. In Chapter 3, "Cultural Perspectives in African Adult Education: Indigenous Ways of Knowing in Lifelong Learning," Ladi Semali analyzes new ways of understanding adult education with respect to African ways of knowing and learning so as to attach it to the world of Africa. Instead of locating adult education on a separate platform of nonformal education, Semali proposes that we need to see it within the context of holistic programs that inform other forms of education and relate it to the lifelong learning of peoples and societies. In addition, Semali suggests that we view and interact with lifelong education as a way of learning and doing that enables local innovation and discovery, which will enhance the specialized capacities of workers and their communities. In Chapter 4, "Mwalimu's Mission: Julius Nyerere as (Adult) Educator and Philosopher of Community Development," Christine Mhina and Ali A. Abdi discuss the contributions of Tanzania's first postcolonial president, Julius Nyerere, who passed away in 1999 but who left a rich legacy of rethinking and re-creating the learning and development contexts of his people. Although Nyerere's programs did not practically succeed as well as he hoped, and were never short of detractors, his understanding of the fundamental need to create an Africanized version of learning possibilities held great promise. With the right elements for their success in place, these programs could have produced more inclusive educational and livelihood prospects for Tanzanians and potentially for others elsewhere in the continent and around the world.

In Chapter 5, "Globalization, Dispossession, and Subaltern Social Movement (SSM) Learning in the South", Dip Kapoor examines development and neoliberal globalization as processes of political-economic and cultural imperialism (successive global capitalist colonizations) and considers their displacing and dispossessing implications for subaltern social groups/ecological ethnicities (indigenous peoples, peasants, landless workers, Dalits/untouchable castes, and rural/forest-based communities subjected to the social relations of domination) in the South. Kapoor also examines the increasing importance of critical/popular adult education and subaltern informal learning in SSMs. Emergent analytical and praxiological dimensions for adult learning (from Kapoor's current research on *Adivasi*/original dweller movements and his decade-long association with *Adivasis* in eastern India) and popular education in SSMs are introduced, while Kapoor also makes the case and offers a tentative schema for disembedding SSMs from their "globalization from below" (new social movements [NSMs]) Euro-American categorization, since their teleologies as well as their historic and contemporary struggles are arguably quite distinct from NSMs or OSMs (old social movements).

In Chapter 6, "Paulo Freire and Adult Education," Peter Mayo selectively locates the work of Freire in its rightful context: adult education and its role in emancipating men and women from submerged consciousness. Mayo, one of the foremost scholars on Freirean and Gramscian philosophies of education (see his excellent book, *Gramsci, Freire and Adult Education*, 1999) starts with the Brazilian thinker's formative years, alludes to how he was exposed to Marxian thinking while in exile in Bolivia and Chile, and relates how Freire uncompromisingly focused on the rights of people vis-à-vis the desires and demands of economic determinism. In addition to the written works, Mayo uses some primary resources (including an interview with Freire's widow Ana Maria Araujo Freire in 2000) to share important snapshots of the life of this brilliant humanist philosopher of education. In Chapter 7, "Freire and Popular Education in Indonesia: INSIST and the Indonesian Volunteers for Social Transformation (Involvement) Program," Muhammad Agus Nuryatno discusses the experiences and educational programs of the Indonesian Society for Social Transformation (INSIST), including a transformative course for nongovernmental organization (NGO) workers and activists. This course, based on Freirean critical pedagogy, is intended to enhance the critical capacities of its recipients. Nuryatno provides details of the program with respect to its participants, the emphasis of instructional programs, and its impact on social change and social development.

In Chapter 8, "Nonformal Education, Economic Growth, and Development: Challenges for Rural Buddhists in Bangladesh," Bijoy Barua examines nonformal/extension education for community and social development in a Buddhist village within the present socioeconomic context of Bangladesh. Based on a qualitative study conducted in the village of Muni, Barua argues that nonformal/extension education, deeply embedded within materialistic and exploitative colonial roots, has adopted the agenda of cultural homogenization within the problematic camouflage of development. Barua points out ways of dealing with this, such as appreciating community-based Buddhist experimental learning as a counterweight to current learning relationships. In Chapter 9, "East Meets West, Dewey Meets Confucius and Mao: A Philosophical Analysis of Adult Education in China," Shibao Guo looks at the contributions of these important thinkers and how the general development of adult education has benefited from their writings, thought systems, and emphasis on learning. He also explores how these different and at times divergent ideological forces coexisted and somehow constructively interacted to shape China's adult education programs into its current context. Guo analyzes their levels of influence and seeks out ways of establishing who had the biggest impact on the ongoing formations and modifications of the system.

In Chapter 10, "The Role of Continuing Education in Zimbabwe," Michael Kariwo discusses the important role continuing education could play in the lives of people in Zimbabwe. Kariwo provides a conceptual analysis of continuing education and relates adult learning possibilities with respect to the country's educational resources and capacities. He also discusses how the development of adult education could help Zimbabwe deal with the many social development problems it is facing, a situation found not only in this country but in many African countries where the promise of postcolonial advancement did not fully materialize. In Chapter 11, "Popular Education and Organized Response to Gold Mining in Ghana," Valerie Kwaipun looks at the different, organized responses of local communities to the rising number of surface-mining activities in western Ghana. In her analysis, Kwaipun provides historical background complemented with some foci on the political economy of Ghana's mining sector. She then discusses the environmental impact of the mining projects in respect to the emergent community responses, especially those undertaken by NGO groups who support the rights of the local communities.

In Chapter 12, "Popular Education, Hegemony, and Street Children in Brazil: Toward an Ethnographic Praxis," Samuel Veissière and Marcelo Diversi situate the world of street children at the critical intersections of adult education and how its attachments to dominant discourses and practices could actually explain the overall context and possibilities. In so doing, Veissière and Diversi present their paradigm of adult education as one that can rely on the knowledge constructions of marginalized populations and consider how it can point out ways of understanding oppression and overcoming it. Fully related to this is the authors' attempt to establish inclusive critical platforms that show the weaknesses inherent in the way programs for street children are designed and implemented. In Chapter 13, "Citizens Educating Themselves: The case of Argentina in the Post–Economic Collapse Era," Luis D'Elia examines the exceptional community-based sociocultural and political conditions that enabled new movement groups (autonomous workers' groups) to devise pragmatic popular responses to deal with Argentina's economic collapse in 2001. D'Elia describes and analyzes how hundreds of thousands decided to organize and, in Latin America's excellent tradition of popular learning, educate themselves instead of waiting for the usually ineffective national institutions to deal with the problems directly.

In Chapter 14, "Freirean Analyses of the Process of Conscientization in the Argentinean Madres Movement," Charlotte Baltodano uses Freirean methodologies as important analytical tools to comprehend the learning dimensions of the Madres movement (mothers of the disappeared) in Argentina during the late 1970s. In discussing the issue, the author examines the role played by emancipatory learning possibilities that helped sustain the movement from the early needs-based action to long-term sustained social movement activism. Based on her own social activism in Latin America, Baltodano makes a case for the importance of learning and the development of critical social consciousness with respect to the Madres movement. Finally, in Chapter 15, "Adult Education and Development in the Caribbean," Jean Walrond provides some background on the context of the area and traces select adult education development programs that have taken place there, and discusses the idea of development in regard to the different social and economic sectors that benefit from available learning possibilities. It is made clear that the development of adult education in the Caribbean evolved from select learning platforms that were modified and improved as needs arose. In addition, Walrond speaks about the influences of some informal education sectors that were also conducive to the advancement of learning possibilities within and around people's lived contexts. The author makes important notations on specific popular education contributions made by significant political leaders, such as Trinidad and Tobago's first postcolonial prime minister, Eric Williams, whose book *Massa Day Done* was seen by many as an important manifesto for upholding the rights of the common man and women.

Together, these chapters represent not only different foci on the expansive theorizations and practices of international adult education, but also the divergent and intellectually evolving interests of the contributors. While some chapters deal with conceptualization more than others, the complementary nature of combining the theoretical and the case study should create a balance in understanding and appreciating all aspects of the adult education platform. We hope that readers will thus rethink ways of explaining global learning contexts and, in the process, reconstitute for themselves and for others the primary right to selectively attach these comments to real and realizable live contexts. Obviously, this is not a complete project and, just as with any new debates that arise in the academic public forum, the main intention is to add something situationally useful to the ongoing and ever changing trajectories of ideas, knowledge, and ways of knowing. Even from the inclusiveness of the historiographical corner, we were not able to add chapters on Oceania and other important places, but we are confident that future works will cover more of our world, and for all pragmatic undertakings, ours should be viewed as a timely introductory work (for us, it is) that expands the forum and gives some deserved voice to the very active and complex adult education contexts that are particular to the so-called developing world. As we have already indicated, we attach important human well-being possibilities to educational programs and, as such, any discussion of learning intersections that concern the lives of the more marginalized should add value to the potential for transforming the lives of people. In sum, we hope this work informs others about such contexts and instigates more debates that might lead to more refined and socioculturally inclusive appreciations of the situation.

References

Abdi, A.A. (2002). *Culture, education and development in South Africa: Historical and contemporary perspectives.* Westport, CT: Bergin & Garvey.

Abdi, A.A. (2006). Culture of education, social development and globalization: Historical and current analyses of Africa. In A. Abdi, P. Puplampu, & G. Dei (Eds.), *African education and globalization: Critical perspectives* (pp. 13–30). Lanham, MD: Rowman & Littlefield.

Achebe, C. (2000). *Home and exile.* New York: Oxford University Press.

Cesaire, A. (2000). *Cahier d'un retour au pays natal.* Columbus, OH: Ohio State University Press. (Original work published 1947)

Coare, P., & Johnston, R. (2003). Introduction. In P. Coare & R. Johnston (Eds.), *Adult learning, citizenship and community voices* (pp. viii-xv). Leicester, UK: NIACE.

Crowther, J., Galloway, V., & Martin, I. (Eds.), (2005). *Popular education: Engaging the academy, international perspectives.* Leicester, UK: NIACE.

Dewey, J. (1926). *Democracy and education.* New York: Collier Macmillan.

Fanon, F. (1968). *The wretched of the earth.* New York: Grove Press.

Fenwick, T. & Tennant, M. (2004). Understanding adult learners. In G. Foley (Ed.), *Dimensions of adult learning: Adult education and training in a global era.* (pp. 55–73)Milton Keys, UK: The Open University Press.

Freire, P. (1985). *Education for critical consciousness.* South Hadley, MA: Bergin & Garvey.

Freire, P. (2000). *Pedagogy of the oppressed.* New York: Continuum. (Original work published 1970)

Freire, P. (2001). *Pedagogy of freedom: Ethics, democracy and civic courage.* Lanham, MD: Rowman & Littlefield.

Harris, E. (1996). Revisioning citizenship for the global village: Implications for adult education. *Convergence, 29*(4), 5–12.

Held, D. (Ed.). (1999). *Global transformations: Politics, economic, culture.* London: Polity.
Jarvis, P. (2004). *Adult education and lifelong learning: Theory and practice.* New York: RoutledgeFalmer.
Johnston, R. (2003). Adult learning and citizenship: Clearing the ground. In P. Coare & R. Johnston (Eds.), *Adult learning, citizenship and community voices* (pp. 3–21). Leicester, UK: NIACE.
Lindeman, E. (1961). *The meaning of adult education.* Montreal: Harvest House
Macaulay, T. (1995). A minute on Indian education. In B. Ashcroft G. Griffiths, & (Eds.), *The post-colonial studies reader* (pp. 428–430). New York: Routledge. (Original work published 1935)
Memmi, A. (1991). *The colonizer and the colonized.* Boston: Beacon Press. (Original work published 1957)
Nandy, A. (1997). Mental colonization. In M. Rahnema & V. Bowtree (Eds.), *The post-development reader* (pp. 168–178). London: Zed Books.
Nyerere, J. (1968). *Freedom and socialism: A selection from writing and speeches, 1965–67.* London: Oxford University Press.
Nyerere, J. (1974). *Man and development.* New York: Oxford University Press.
Pöggeler, F. (2002). Globalization through the transfer of adult education between cultures: Perspectives on past and present. In B. Hake, B. van Gent, & J. Katus (Eds.), *Adult education and globalization: Past and present* (pp. 107–115). New York: Peter Lang.
Rahnema, M. (1997). Introduction. In M. Rahnema & V. Bowtree (Eds.), *The post-development reader* (pp. 377–404). London: Zed Books.
Rodney, W. (1982). *How Europe underdeveloped Africa.* Washington, DC: Howard University Press.
Saha, L. (2006). Citizenship and participation in lifelong learning. In A. Antikainen, P. Harinen, & C. Torres (Eds.), *In from the margins: Adult education, work and society* (pp. 81–87). Rotterdam, Holland: Sense Publishers.
Stiglitz, J. (2002). *Globalization and its discontents.* New York: W.W. Norton.
Sutton, P. (1981). *Forged from the love of liberty: Selected speeches of Dr. Eric Williams.* Trinidad: Longman Caribbean.
Tikly, L. (2001). Post colonialism and comparative education research. I.K. Watson (Ed.), *Doing comparative education research* (pp 245–264). London: Symposium Books.
United Nations Development Program (2003). *Human development report.* New York: Oxford University Press.
van Gent, B. (2002). Lessons in unity: Cultural education and social cohesion. In B. Hake, B. van Gent, & J. Katus (Eds.), *Adult education and globalization: Past and present* (pp. 99–106). New York: Peter Lang.
Wallerstein, I. (2004). *World Systems analysis. An introduction.* Durham, NC: Duke United Press.
wa Thiongo, N. (1986). *Decolonising the mind: The politics of language in African literature.* London: James Curry.

CHAPTER 2

Globalization and Adult Education in the South

Edward Shizha and Ali A. Abdi

In today's deceptive and selectively globalized world, education is one of the most crucial and critical resources for acquiring the skills and capabilities needed to function in competitive capitalist job markets. Global capitalism and its disequalizing dynamics (Alam, 2003) reflect major structural imbalances in the distribution of global resources. Deficits in international trade and investment have resulted in limited employment opportunities in the South (which we will interchangeably call low-income countries). Furthermore, professionals and skilled workers, the core human resource for economic and social development, are leaving in droves for better life opportunities in Europe and North America; those without the requisite internationally recognized skills are using adult education to reskill themselves or upgrade their academic credentials to enter these lucrative labor markets.

While globalization camouflages itself as a development philosophy that promises economic growth through neoliberalism, for most countries in the South, the adoption of neoliberal trade and investment policies has caused extensive suffering instead of improving economic growth and quality of life. Starting in the 1980s, the open-door economic regimes imposed by the World Bank (WB) and International Monetary Fund (IMF), the core capitalist financial institutions, have produced no economic miracles but instead have reduced state ownership and involvement in industry and encouraged cost-sharing in social services such as education and health. In effect, these neoliberal policies have pauperized the majority of citizens (Shizha, 2006a).

In this chapter, we explore the impact of globalization on adult education and employability in the countries of the South. We also discuss how professionals and skilled individuals are seeking skills and educational credentials through adult education and communication technologies to enhance their employability in developed or high-income countries, thus contributing to brain drain and underdevelopment in their home countries.

Neoliberal Economic Globalization and Adult Education in the South

There is a connection between globalization/neoliberalism and education in developing countries. The relationship involves a politico-economic discourse of power and control. The North controls globalization and manipulates global economics. Neoliberalism can be contextualized within the globalization debate. Globalization is a multilevel phenomenon that is both ideological and economic (Stromquist, 2002). It is marked by dramatic developments in communication technologies that have compressed time and space (Giddens, 1990) which, in turn, have caused significant changes in the economic and social landscapes among nation-states (Burbules & Torres, 2000). These relations are marked by differentiated power dynamics that prioritize the politico-economic interests of countries in the North. While political globalization is comprised of, at least partially, economic and social involvement by the state, economic globalization emphasizes open marketization of production and free movement of financial capital (Puplampu, 2005). From a postcolonial perspective, globalization is not a new phenomenon, but a continuing and expansive project of neocolonialism or recolonization of the South. For Africa, Asia, the Caribbean, and Latin America, globalization is an extension of colonial imperialism camouflaged in terms such as *market forces*, *poverty reduction*, and *liberalization* rebranded from outside developing nations. However, rebranding old processes cannot bridge the economic imbalance between the North and the South. Neither can it reduce poverty and the gap between the rich and poor nations as speechified in the annual G8 leaders' rhetoric. In effect, globalization has diminished the prospects of instituting equity and social development around the world (Mkandawire, 2002; Zeleza, 2004). The grinding poverty in the South, especially sub-Saharan Africa (Abdi, Puplampu, & Dei, 2006), compared to the dramatic economic development in the North (Stromquist, 2002), displays a poignant economic situation resulting from neoliberal policies. Given the prevailing economic inequalities between the North and the South, it is not surprising that several African and Caribbean countries are at the lower end of the United Nations Development Programs (UNDP, 2001) annual Human Development Index. In essence, we view globalization as an underdeveloping process.

Globalization and dehumanization are two faces of contemporary capitalism and social fascism (Dale & Robertson, 2004) which have caused considerable social dislocation. The impact of globalization, which was formally framed in modernity, has left a debilitating mark on social and economic transformation in the South. Although poverty in Africa, Asia, Latin America, and the Caribbean is historical and a product of capitalist systems entrenched in colonial legacies, the liberal discourse of "free" trade has added nothing to equalize and balance trade relationships between the North and South. What is undoubtedly observable and obvious is the intensification of the European and American political and economic exploitation of low-income nations through the embodiment, reinvention, and reinforcement of the exploitative capitalist economic system.

Describing the African context, Dei and Asgharzadeh (2006) state that global capitalism continues to challenge the social and cultural existence of African people, and is to blame for the economic crisis in most nations in the South. The reinvention of economic colonialism threatens the unique localized ways of creating self-sustaining production and survival, thus increasing the need for acquiring new skills among adult learners. The necessity for skills acquisition and national development draws attention to the failure of adult educational programs and policies to address the economic and social malaise in developing nations (Puplampu, 2005). In developing countries, universities are often institutions for the development of high-skilled manpower, technology transfer, and generation of knowledge (Kariwo, 2007); however, inadequate funding undermines their capacity to contribute to national and economic development. This funding disparity limits the institution's ability to provide their students with knowledge and skills relevant to the local and global economy. Higher education in the South has therefore suffered due to emphasis placed on "managerial efficiency and quantitative measurements at the expense of qualitative aspects of education" (Zeleza, 2004, p. 53).

Global neoliberal policies have forced compliant governments to cut financial support for basic education and to refocus educational policies and expansion on higher education, subject to corporate ideology (Kariwo, 2007). Consequently, governments and universities are redefining educational policies and restructuring postsecondary education along entrepreneurial lines with limited public funding. The greater commoditization of education into marketable merchandise has turned educational institutions into essentially commercial organizations (Jarvis, 1995, p. 221). The assumption is that students, as willing buyers, will grab opportunities to acquire skills for the competitive global labor market. These educational reforms are regarded as an investment for individuals rather than a public good for social transformation. Costs and benefits are evaluated according to market principles, thus limiting intervention policy instruments available to the state to provide basic needs of citizens. As a result of these neoliberal education policies, low-income countries are producing fewer professionals, scientists, technical experts, and educated adults required for local economic and social development.

The World Bank and Education

The World Bank and IMF ideology is imposed in the South as structural adjustment programs (SAPs) to promote market forces in economic development. Loans from the World Bank and IMF are invariably made conditional on the implementation of monetary deregulation, trade liberalization and liberalization of capital markets, privatization, and enforced fiscal constraints (Shizha, 2006a). SAPs have also forced compliant governments to reduce expenditures in social services, thus increasing poverty among citizens contrary to the World Bank's claim that SAPs are meant to increase economic production and reduce poverty in low-income countries.

The criticism against SAPs led the World Bank to launch the Poverty Reduction Strategy Initiative, linked to the Poverty Reduction Strategy Papers (PRSP; World Bank, 2003), which encourage countries to develop poverty reduction strategies that would be financed by the World Bank and IMF. However, the World Bank still enforces the same austerity conditions enshrined in SAPs. The policies of the PRSP reinforce the unequal global socioeconomic relations between the North and the South (Shizha, 2006a). Export-oriented growth and market-induced reform models, as imposed by the WB and IMF, only serve the interests of international monetary agencies and the national bourgeoisie. In effect, the IMF and World Bank demand that poor nations lower the standard of living of their people in order to maintain debt servicing, while debt and so-called open markets contribute to poverty and underdevelopment.

In addition to SAPs that are politically difficult for governments because they negatively impact expenditure on education, neoliberal policies have adversely reshaped, redirected, and redefined educational policies in Africa (Shizha, 2006a) and Latin America (Carnoy & Torres, 1992). According to the World Bank (2001), their education policies promote and emphasize the rate of social return. In the early 1960s, the World Bank's policy on higher education in Africa was human resource development and the Bank provided significant financial resources to universities. By the 1970s, however, the Bank's policy shifted to focus on the rate of return analysis (Samoff & Carrol, 2004). Their main—misleading—argument was that society derives higher social rates of return from investment in basic education and not higher education. Governments were, therefore, required to reduce expenditure on higher education as a condition for loans (Samoff & Carrol, 2004). This led to the reduction of student numbers in several African universities. Thereafter, ongoing educational reforms in many developing countries under the auspices of the World Bank meant that large segments of local populations were left without higher education as the cost of schooling became unbearable (Shizha, 2006a).

Most African countries witnessed significant investments in education in the first two decades of independence. Although Africa continued to lag behind other regions in the South, primary and postsecondary education expanded at an impressive rate. Between 1965 and 1985, primary school enrollments rose from 41 percent of the eligible population to 68 percent, while enrollments in postsecondary institutions increased at a similar pace (Mkandawire and Soludo 1999, p. 16). However, with the advent of SAPs, governments abandoned progressive educational programs necessary for socioeconomic transformation (Mkosi, 2006), resulting in a reduction in school enrolments and a marked increase in dropout rates among students (Shizha, 2006a). In Latin America, educational levels across the region have declined since the 1990s. In Argentina, for example, 14 percent of those over 15 years old are illiterate, compared with 7.4 percent in 1970 and 6.1 percent in 1980 (Puiggros, 1996).

In all nations, as resources become scarce, there is a tendency for political intervention in the resource allocation process (Kariwo, 2007). In developing nations, the pathological allocation of domestic public resources to the education sector on the advice of what the global financial institutions perceive to be economic rationale has depleted educational provision and quality leading to a slowdown in human resource development (Shizha, 2006a). Financial decisions affecting educational policy no longer focus on improving educational quality, but instead on meeting the financial goals of SAPs (Carnoy & Torres, 1992). The human element is ignored in favor of budgetary austerity. While social demands have increased, public resources have not. Big cutbacks in public funding combined with an increase in poverty have reduced possibilities for improving workforce development and indigenous advancement in science and technology.

In Africa, as part of instituting fiscal constraints, cost sharing in education was introduced, forcing parents to directly contribute toward their children's educational cost (Shizha, 2006a). The cost-benefit evaluation of output became a way of controlling educational demand, with the goal being to reduce government investment in public education, hence marginalizing access from poor families. Cost sharing in education is linked to the commoditization and marketization of education discussed earlier in this chapter. These policies have led to the "massification of private education" for those who can afford to pay. The "fundamentalism of finance capital and the elevation of the market to the status of a universal deity" (wa Thiong'o, 2005, p. 155) are evident in governments' political and economic decisions when it comes to financing universities in Africa, Asia, and Latin America. The retreat by governments from their social obligations to finance higher education has been a severe blow to the South, just when the role of knowledge in national development was becoming accentuated.

From Basic Adult Education to Economic-Focused Education

There is wide agreement among policy makers inside and outside of developing countries that the overarching education priority in most countries must be to attain universal education. Yet education for All, the goal set at the World Conference held in Jomtien, Thailand, in 1990 was never achieved. At a follow-up conference in Dakar a decade later, the World Education Forum pronounced that goal anew, but for most developing countries this proposal has been overshadowed by economic problems. The educational reform processes (Muzvidziwa & Seotsanyana, 2002) and social programs, including free basic education, to that were begun by newly independent developing nations to bridge the educational gap created by colonial governments have since faltered while economic growth remains stunted.

The 1990s saw the development of a culture of adult education in the form of "lifelong education" as a key objective of governments concerned with balancing issues of economic development and competitiveness in global markets (Prinsloo, 2005). Lifelong learning was a policy shift from basic education that provided literacy to adults to education that focused on widening participation at the tertiary level. It was a change from the social agenda of pre-1990s to an economic development agenda (Muzvidziwa & Seotsanyana, 2002). Taking their cue from neoliberalism, adult education practices were divorced from education for social and personal transformation in favor of hyper-individualism (Giroux, 2004). Thus, instead of consolidating collaboration and social interdependence, the new policies promoted consumerism, individualism, and competitiveness in the information society. Because effective participation in a global environment depends on the ability and skills necessary to function within the information/knowledge society (Castells, 2001), knowledge and information have become central to globalization and national development.

A core component of the knowledge economy is human resources. Without human resources, nothing works. Human resources require not just technical skills in a minority, but a broad level of education in the population at large. Based on the belief that primary and secondary schooling are more important than tertiary education for poverty reduction, higher education was previously relatively neglected by most African, Asian, and Latin American governments. According to the Education for All Report, by 2004 many countries in the North had significantly universalized basic education, while within South and West Asia the rate was 92 percent. This was significantly higher than rates in sub-Saharan Africa, which stood at 73 percent, but still below those of East Asia (106 percent) and Latin America and the Caribbean (110 percent (UNESCO, 2005). Tertiary education, however, is still a luxury in all developing countries and limited to the few who can afford it. Africa records the lowest enrolment rates in higher education in the world (Sawyerr, 2002). Global capitalism negatively affects Africa's ability to compete in international trade, and the flow of its goods to international markets, weakening Africans governments, financial resources to fund higher education.

With the advent of globalization, governments have focused more and more on tertiary education as a means for imparting future workers with the skills necessary for economic development. Knowledge-based competition within a globalizing economy is prompting a fresh consideration of the role of higher education in development (Bloom, Canning, & Chan, 2006). Refocusing the role of higher education to economic growth is not an original discourse; the development perspective is entrenched in the modernization theory of the 1960s and 1970s. Higher education has been given much of the burden for transforming the African, Asian, Latin American, and Caribbean economies. This transformation requires universities to provide expertise, experience, and leadership in workforce development and to generate transferable ideas and technologies from research to commercial applications (Sawyerr, 2002). Despite these noble requirements, developing countries are not committed financially and politically to channel adequate resources to higher education to help the future workforce gain a competitive edge in today's increasingly competitive economy. The funding of university education remains inadequate for the needs of the knowledge society (Mkosi, 2006). Put simply, the intended massification of higher education for economic development cannot be achieved without adequate resources to pay for the education.

Knowledge Commoditization, Credentialism, and Reskilling

Expectations of what adult education should achieve range from fostering sustainable development to promoting democracy, justice, equity, and social development (UNESCO, 1997). Central to debates on adult education are human resource development and retraining programs (Finger & Asun, 2001). Advancements in information and communication technology (ICT) have detraditionalized education systems and training programs. To attain the "good life" from the labor market, individuals are now redefining and reconstructing their personal, working, and public lives (Glastra, Hake, & Schedler, 2004) through retraining and reengaging in adult education. Many individuals seek adult education for the acquisition of occupational skills. For these people, continued educational plans originate more or less from these instrumental considerations (Massey et al., 1998). New professional qualifications and credentials are likely to increase job opportunities due to the changing work environment that requires vocational and scientific skills and superior credentials. As a result, increased participation in higher education has been recorded over the years in many developing countries.

In Africa, the number of students attending university has risen dramatically in the last three decades. From an estimated 600,000 students in 1980, the number increased to 1,750,000 by 1995 (Sawyerr, 2002, p. 13). Nigerian universities recorded a whopping increase from 176,000 in 1989/1990 to 376,000 by 2000 (Sawyerr, 2002, p. 14). While we may argue that the rise in the demand for higher education is justifiable considering the need for skilled labor, we should not be blind to the way education is being commercialized. The economic dimension of adult education is highlighted by the expansion of distance education and the presence of imported ICT, widening access to adult education. Although imported technology offers global educational possibilities, its ideological and contextual inappropriateness can alienate students. Knowledge transmitted through ICT becomes exteriorized from students in developing countries. Context embeds knowledge in sociocultural experiences. As Giroux (2001) asserts, "unless the pedagogical conditions exist to connect forms of knowledge to the lived experiences, histories, and cultures of the students we engage, such knowledge is reified or "deposited" in a Freirian sense, through the transmission models that ignore the context in which knowledge is produced." (p.11). While the expansion in knowledge dissemination is not bad per se, its form should be culturally contextualized especially in the current neoliberal era in which commercialization dictates educational administration while disregarding the social roles and responsibilities of public education. Universities are more and more concerned about accounting than accountability (Tettey, 2006) and satisfying the educational business enterprise (Altbach, 2002).

Because of the high demand for knowledge relevant to economic demands, the rush for adult education has led to the commercialization and commoditization of knowledge. The rush to translate knowledge into marketable information (Lyotard, 1984) has alienated students from the social realities that exist in their cultural milieu. Community knowledge is viewed as lacking utilitarian value and hence not worthy of acquiring. Individuals seeking new skills no longer see education as a social activity for the common good, but as an individual enterprise for socioeconomic mobility, consequently undermining social and cultural relationships.

Arguably, in the global environment, education for economic reproduction has displaced education for social justice. Globalization is forcing adults from developing nations to scramble for instrumental knowledge, credentials, and vocational skills that can heighten their employability. Credentialism (Collins, 1979) has become an emerging reality. New types of credentials incorporating new values and skills are being sought because individuals have lost confidence in the old types that restrict them to outdated work ethics and attitudes. The quest for credentials has led to educational consumerism. Students are willing to pay heavily for commoditized knowledge, no matter how alienating it is to local conditions, as long as it opens doors to new credentials. We believe that consumerism and commercialization are no longer the exclusive terrain of business people, but have also become an attractive proposition for many academic institutions as the global decline in state funding for higher education takes root (Tettey, 2006). In that context, students in Africa and in other low-income nations are reduced to "consumers" that universities "market" their "services and products" to, and the desired outcomes are the capitalist attitudes required by employers and what the "international market demands" in skills. (Carnoy, 2000)

Migration and the Brain Drain

The migration of highly qualified professionals from developing countries is an extremely complex problem for the South. The decision to migrate is often related to aspects of professional development and improvement in quality of life (Ferro, 2006). The rich capitalist economies attract highly skilled people pushed out of their countries by poverty, declining economies, lack of employment opportunities, and poor living standards. Poverty in developing nations is a historical and political phenomenon (aptly discussed in Walter Rodney's 1972 book, *How Europe Underdeveloped Africa*). The exploitation of resources and the undervaluation of currency and exports by Europe created a dependency relationship that disadvantaged and impoverished Africa and other former European colonies. As discussed earlier in this chapter, neoliberalism has extended this dependency syndrome and exacerbated poverty and underdevelopment in developing nations (Shizha, 2006a; Mkosi, 2006). Ironically, the very professionals from the South who were supposed to be involved in the economic transformation and renaissance of their countries are instead trekking in the thousands to the North in pursuit of better life.

On a global level, the free movement and interaction of highly skilled people is a positive thing, especially for the destination country that benefits from the brain gain. But the cost to the home countries of losing their professionals is incalculable in terms of both development opportunities and loss of investment. For instance, about 100,000 Indian professionals, primarily in the computer industry, migrate each year to the United States (Human Development Report, 2001), while Africa loses more than 20,000 skilled personnel annually to the developed world (Sriskandarajah, 2005). Consequently, this implies that there are 20,000 fewer skilled people in Africa to deliver key public services (such as education and health) and to drive economic growth and national development. Recent OECD data suggest that around two-thirds of all highly skilled workers from Guyana, Jamaica, Haiti, Trinidad and Tobago, and Fiji have left these countries (Sriskandarajah, 2005) for greener pastures. The sheer volume of emigration suggests that any possible positive returns to the home countries may be outweighed by lethargic and pathological impacts on national development.

To curb the brain drain, developing nations should ignore the World Bank's policy prescription and reinvest in both economic and social services in order to improve the quality of their people's lives. Investment should be directed toward the needs of those living in the informal subsistence economy through community-centered education (Bhola, 1999). The influence of globalization should not be to impose political and economic policies from the outside, but to readjust and incorporate homegrown policies in order to develop a country's capacity to respond to the expectations of the global economy. Attainment of educational and professional qualifications should be driven by the desire to satisfy the multiplicity of economic roles that local labor demands and avoid hegemonic forces that lack focus on the necessary productive skills needed for overall community development.

Globalization, Adult Education Information, and Communication Technologies

Globalization is dependent on the role of information and communication technologies in facilitating instant access to information without regard to spatial dimensions and time horizons (Glastra et al., 2004). This process has led to the rise of distance education and virtual education, such as the African Virtual University in Kenya. The provision of education is no longer left to what is local and familiar. While universities are still recognized as critical to the knowledge society, they are no longer considered as the sole agency for the production, use, and dissemination of knowledge (Prinsloo, 2005). Information technology and virtual/cyber universities are gradually being introduced in the South to cater for prospective students who fail to get places at traditional universities. The introduction of information technology into the educational system is partly an attempt to expand the quantity of educational provision at lower cost through distance education and partly to deliver higher quality education through computer-assisted instruction via the Internet (Carnoy & Rhoten, 2002).

Wholly Internet-based electronic learning has recently captured the attention of education policy makers. The spread of the Internet has made online learning very common. Although some countries in the South have joined the ICT revolution, they face challenges in acquiring the technology. For example, in Africa, despite the 52.1 percent growth in, availability of the Internet between 2000 and 2002, its usage was limited to 2.1 percent of users and it has penetrated merely 0.8 percent of household among the population of 859 million (LaRocque & Latham, 2003). So the possibility of mass effective use of ICT in adult education in these countries is not realistic due to underfunded education programs. Furthermore, where technology exists, blind use of imported technology without contesting and interrogating its relevance and appropriateness to African, Asian, or Latin American economic developmental situations may lead to misrepresentation of national needs and thus to "an epistemology of the colonized" (Dei & Asgharzadeh, 2001, p. 300). Unless technology is harnessed to produce knowledge networks that are responsive to social and economic needs of the countries where that technology is adopted, the knowledge gained may become an unproductive intellectual pursuit in worthless credentials.

Cultural Globalization, Cultural Identities, and Indigenization

The debate on globalization and education must involve a discussion of cultural transformations. As a consequence of globalization, cultural dynamism has occurred both in the North and South. The multicultural lifestyles that people are experiencing and their responses to the physical and social environments are an indication of the dynamic nature of culture. National cultures are subjected to both internal and external forces (Abdi, 2006) via transnational migration and the pervasive nature of technology that is exposing people to diverse cultures. Regrettably, the hegemonic cultural colonialism or imperialism (Memmi, 1991) imposed during the nineteenth century on the people of Africa, Asia, Latin America, and the Caribbean, as well as the emergence of cultural globalization, have had a disruptive effect on the lives of indigenous people. Cultural imperialism, labeled as a form of enlightenment, destroyed indigenous cultural lives that were harmonious and responsive to the economic, social, and political situations that existed within their societies.

We argue that cultural globalization is framed in a discourse of politics of identity. In many societies, political and cultural consciousness have heightened collective identities in order to preserve cultural uniqueness, and cultural globalization is seen as a threat to local distinctiveness and self-determination. While some critics view globalization as a form of hybridization, implying the blending of cultures to form a homogenized way of life (Pieterse, 2004), others see it as resulting in differentialism or "a mosaic of immutably different cultures and civilizations" (Beynon & Dunkerley, 2000, p. 121). In a diverse and differentiated world, cultural homogenization is not attainable. People from different cultural communities will always attempt to protect their ethnic identities. Identity politics or ethnification is a counterreaction to individualism introduced by globalization (Abdi, 2006). It is also a defense mechanism used by communities against foreign dominance (Mkosi, 2006; Shizha, 2006a). By and large, it can also be viewed as competition for scarce resources between groups (Shizha, 2005), or a strategy of exclusion (Aikenhead, 2000). Identity politics always entails competition over scarce resources and recognition.

With globalization comes a growing awareness at the local level of the need to protect cultural identity and cultural diversity. The diversity of race, language, class, and religion that exists in today's world cannot be ignored. Even with the pervasive nature of globalization, cultural identities are not easy to eliminate. People will adapt their ways of life, but those adaptations are guided and controlled by the values and norms that have been accumulated over generations. Education systems reflect culture, and culture-based education validates the values, worldviews, and languages of a particular society (Shizha, 2006b). Adult education also is influenced by and has influence on the society in which it is practiced. In precolonial Africa, Asia, and Latin America, education was in harmony with culture. It developed at a pace that was responsive to the needs imposed by the human and physical environment of the day (Abdi, 2006). Education focused on validating people's experiences and linked people's lives to their indigenous knowledges, which in turn were linked to politics of identity and development and poverty alleviation (Dei & Asgharzadeh, 2006). All life possibilities, including educational systems, were more pragmatic and more attached to the life of the indigenous people (Abdi, 2006). Within the field of education across the South, basic education was linked to attaining survival skills. Education was a community and participatory enterprise involving adults and the young, and knowledge was rigorously created and transformed to adapt to the ever-changing socioeconomic needs of the people (Appleton, Fernandez, Hill, & Quiroz, 1995; Shizha, 2006b).

Despite the influence of Western science that undervalues indigenous knowledge, local knowledge continues to be observed in indigenous communities that have resisted extermination. These communities are resilient because they satisfy collective human needs rather than the individual consumption that global knowledge promotes. Local knowledge signifies the power of the intellectual agency of local peoples (Dei & Asgharzadeh, 2006). Globalization has not weakened the resilience of indigenous people to preserve their collective indigenous epistemology, and historical memories and practices that address their holistic needs.

While indigenous knowledge systems are communally created and cater to communal needs, global knowledge systems are individually created and protected by capitalist intellectual property rights (IPRs). Because Western science and technology are embedded in market ideology, the IPRs protect primarily individual and corporate interests within the profit-motivated system. In Africa, Asia, and Latin America, indigenous people have not been accorded IPRs and their collective responsibilities to protect their plant genetic materials have been disrupted and ignored by the colonial and imperial appropriations that began centuries ago.

In the current globalization era, natural resource exploitation and industrial growth are disrupting biodiversity in the ecosystem. The widespread introduction of exotic species to non-native areas is a potent threat to biodiversity. The commercialization of endangered species (Dei & Asgharzadeh, 2006) has resulted in deforestation, soil erosion, and water and air pollution through uncontrolled exploitation and use of forests, land, chemicals, and dispersal of waste material (Appleton et al., 1995). Globalization has thus annihilated indigenous livelihoods and communities and continues to rob indigenous populations of their local knowledge systems through uncontrolled exploitation without the consent of the people. Local knowledge is embedded in an awareness of the collective needs of the community (Aikenhead, 2000), hence its relevance and appropriateness to socioeconomic development. Therefore, full recognition of local knowledge systems is central to sustainable and equitable development. The capacity of indigenous systems to integrate multiple disciplines such as agroforestry, animal husbandry, biodiversity, crop production and preservation, and conservation demonstrates higher levels of efficiency, effectiveness, adaptability, and sustainability than many conventional technologies (Shizha, 2005). It is commendable that international agencies, such as the United Nations, have begun to recognize the importance of preserving and promoting indigenous knowledge and technologies (UNESCO, 2005).

Summary

We have argued and illustrated how globalization is influencing the expansion of adult education in developing countries. Internationally recognized skills and competences have heightened the demand for reskilling and skills upgrading. As such, in many countries in the South, adults are seeking post-literacy and continuing education not only for skills advancement, but also for economic empowerment and consciousness raising (Freire, 1985) in order to survive in the global capitalist labor market and economy. The desire for integration into the global economy is high among individuals from developing countries that cannot provide sufficient rewards and comfortable lifestyles. Due to globalization, adult education has changed from a community-based cultural and social enterprise to an individual and egoistic project. Altruistic adult learning has been overtaken by individual indulgence at the expense of community and national development. Although new educational and economic institutions that are "endowed with a pervasive cultural and economic power" (Federighi, 1997, p. 4) are emerging in the South, the international training system with its highly developed technology and economy which function to socialize international workers into global laborers is attracting a large number of professionals from the South, leading to the brain drain.

We have considered here the intersections of historical and contemporary globalization, and the emerging themes of adult education, and appreciate the evolving perspectives as occasioned by global, national, and personal needs and relationships. If globalization was once perceived via a mainly economic discourse, by now it should be clear that the project of globalization touches almost all aspects of people in almost all countries of the world. As we have discussed in this chapter, the effects of globalization have been good for most of the North, but not so good (actually selectively problematic) for the non-elite segments of Southern populations. Any discussion of adult education in these multi-modernist times must be rooted in these realities and should aim for a type of analysis that is as topically complete and analytically inclusive as possible. No single chapter can address all the current pertinent ideas and information that would do justice to the two main topics under consideration. As such, we have been expectedly selective; we have introduced some theoretical notations, analyzed the very active space between globalization and adult education, and examined the role of information technologies in either facilitating or expanding the possibilities of adult education. In the end, we hope our observations will entice others to further discuss and review, even to recast, some or all of our assumptions.

References

Abdi, Ali A. (2006). Culture of education, social development and globalization: Historical and current analyses of Africa. In A. Abdi, K.P. Puplampu & G.J.S. Dei (Eds.), *African education and globalization: Critical perspectives.* (pp. 13–30). Lanham, MD: Rowman & Littlefield.

Abdi, Ali A., Puplampu, K., & Dei, G. (2006). *African education and globalization: Critical perspectives.* Lanham, MD: Rowman & Littlefield.

Aikenhead, G.S. (2000). Renegotiating the culture of school science. In R. Millar, J. Leach, & J. Osborne (Eds.), *Improving science education: The contribution of research* (pp. 245–264). Buckingham, UK: Open University Press.
Alam, M.S. (2003). *Pauperizing the periphery: Two decades of neoliberal policies.* Accessed October 2007 from http://www.twf.org/News/Y2003/0610-Pauper.html
Altbach, P. (2002). Knowledge and education as international commodities: The collapse of the common good. *Current issues in Catholic Higher Education,* 22, 55–60.
Appleton, H., Fernandez, M.E.; Hill, C.L.M., & Quiroz, C. (1995). Claiming and using indigenous knowledge. *Missing links: Gender equity in science and technology for development.* Ottawa: IDRC.
Beynon, J., & Dunkerley, D. (Eds.). (2000). *Globalization: The reader.* New York: Routledge.
Bhola, H.S. (1999). Equivalent curriculum construction as situated discourse: A case in the context of adult education in Namibia. *Curriculum Inquiry, 29*(4), 459–484.
Bloom, D., Canning, D., & Chan, K. (2006). *Higher education and economic development in Africa.* World Bank website. Accessed from November 2007 http://www.worldbank.org/afr/teia/pdfs/Higher_Education_Econ_Dev.pdf
Burbules, N., & Torres, C.A. (Eds.). (2000). *Globalization and education: Critical perspectives.* New York: Routledge.
Carnoy, M. (2000). *Sustaining the new economy: Work, family and community in the information age.* Cambridge, MA: Harvard University Press.
Carnoy, M., & Rhoten, D. (2002). What does globalization mean for educational change? A comparative approach. *Comparative Education Review, 46,* 1–9.
Carnoy, M., & Torres, C.A. (1992). Educational change and structural adjustment: A case study of Costa Rica. *Working Documents of the Operational Policy and Sector Analysis Division.* Paris: UNESCO.
Castells, M. (2001). *The internet galaxy: Reflections on the internet, business and society.* Oxford: Blackwell.
Collins, R. (1979). *The credential society.* New York: Academic Press.
Dale, R., & Robertson, S. (2004). Interview with Robert Cox. *Globalisation, Societies and Education, 2*(2), 147–160.
Dei, G.S.J., & Asgharzadeh, A. (2001). The power of social theory: The anti-colonial discursive framework. *Journal of Educational Thought, 35*(3), 297–323.
Dei, G.S.J., & Asgharzadeh, A. (2006). Indigenous knowledges and globalization: An African perspective. In A. Abdi & A. Cleghorn (Eds.), *Issues in African education: Sociological perspectives* (pp. 53–78). New York: Palgrave Macmillan.
Federighi, P. (1997). *Glossary of adult education in Europe.* Hamburg: UNESCO Institute for Education.
Ferro, A. (2006). Desired mobility or satisfied immobility? Migratory aspirations among knowledge workers. *Journal of Education and Work, 19*(2), 171–200.
Finger, M., & Asun, J.M. (2001). *Adult education at the crossroads: Learning our way out.* London: Zed Books.
Freire, P. (1985). *The politics of education: Culture, power, and liberation.* South Hadley, MA: Bergin & Garvey.
Giddens, A. (1990). *Consequences of modernity.* Cambridge: Polity.

Giroux, H. (2001). Pedagogy of the depressed: Beyond the new politics of cynicism. *College Literature, 28*(3), 1–13.

Giroux, H. (2004). Public pedagogy and the politics of neo-liberalism: Making the political more liberal. *Policy Future in Education, 2*(3&4), 494–503.

Glastra, F., Hake, B.J., & Schedler, P. (2004). Lifelong learning as transitional learning. *Adult Education Quarterly, 54*(4), 291–307.

Human Development Report. (2001). *Brain drain costs developing countries billions.* New York: United Nations Human Development Program.

Jarvis, P. (1995). *Adult and continuing education: Theory and practice.* Routledge: London & New York.

Kariwo, M.T. (2007). Widening access in higher education in Zimbabwe. *Higher Education Policy, 20*(1), 45–59.

LaRocque, N., & Latham, M. (2003). *The promise of e-learning in Africa: The potential for public-private partnerships.* Armonk, NY: IBM Endowment for the Business of Government.

Lyotard, J. F. (1984). *The postmodern condition: A report on knowledge* (G. Bennington & B. Massumi, Trans.). Minneapolis: University of Minnesota Press.

Massey, D.S., Arango, J., Hugo, G., Kouaouci, A., Pellegrino, A., & Taylor, J.E. (1998). *Worlds in motion: Understanding international migration at the end of the millennium.* Oxford: Clarendon Press.

Memmi, A. (1991). *The colonizer and the colonized.* Boston: Beacon Press.

Mkandawire, T. (2002). Globalisation, equity and social development. *African Sociological Review, 6*(1), 115–137.

Mkandawire, T., & Soludo, C.C. (1999). *Our continent, our future: African perspectives on structural adjustment.* Trenton, NJ: Africa World Press.

Mkosi, N. (2006). International financial institutions and education in South Africa: A critical discussion. In A. Abdi, K. Puplampu, & G. Dei (Eds.), *African education and globalization: Critical perspectives* (pp. 151–164). Lanham, MD: Lexington.

Muzvidziwa, V.N., & Seotsanyana, M. (2002). Continuity, change and growth: Lesotho's education system. *Radical Pedagogy, 4*(2). Accessed from October 2007 http://radicalpedagogy.icaap.org/content/issue4_2/01_muzvidziwa.html

Pieterse, J.N. (2004). *Globalization and culture: Global melange.* Lanham, MD: Rowman & Littlefield.

Prinsloo, R.C. (2005). *Making education responsive to prior learning: From Bologna to Bergen and beyond.* Keynote address prepared for the 29th EUCEN Conference, April 28–30.

Puiggros, A. (1996). World Bank education policy: Market liberalism meets ideological conservatism. *NACLA Report on the Americas,* May/June.

Puplampu, K.P. (2005). National "development" and African universities: A theoretical and sociological analysis. In A. A. Abdi & A. Cleghorn (Eds.), *Issues in African education: Sociological perspectives* (pp. 43–62). New York: Palgrave Macmillan.

Rodney, W. (1972). *How Europe underdeveloped Africa.* London: L'Overture Publications.

Samoff, J., & Carrol, B. (2004). *Conditions, coalitions and influence: The World Bank and higher education in Africa.* Paper presented at the annual conference of the Comparative and International Education Society.

Sawyerr, A. (2002). *Challenges facing African universities: Selected issues.* Accessed from September 2007 http://www.aau.org/english/documents/asa-challenges.pdf

Shizha, E. (2005). Reclaiming our memories: The education dilemma in postcolonial African school curricula. In A.A. Abdi & A. Cleghorn (Eds.), *Issues in African education: Sociological perspectives* (pp. 65–83). New York: Palgrave Macmillan.

Shizha, E. (2006a). Continuity or discontinuity in educational equity: Contradictions in structural adjustment programs in Zimbabwe. In A. Abdi, K. Puplampu, & G.J.S. Dei (Eds.), *African education and globalization: Critical perspectives* (pp. 187–210). Lanham, MD: Lexington.

Shizha, E. (2006b). Legitimizing indigenous knowledge in Zimbabwe: A theoretical analysis of postcolonial school knowledge and its colonial legacy. *Journal of Contemporary Issues in Education, 1*(1), 20–35.

Sriskandarajah, D. (2005). Reassessing the impacts of brain drain on developing countries. *Migration information source: Fresh thought, authoritative data, global reach.* Washington: Migration Policy Institute.

Stromquist, N.P. (2002). *Education in a globalized world: The connectivity of economic power, technology, and knowledge.* Lanham, MD: Rowman & Littlefield.

Tettey, W.J. (2006). Globalization, information technologies, and higher education in Africa: Implications of the market agenda. In A. Abdi, K. Puplampu, & G.J.S. Dei (Eds.), *African education and globalization: Critical perspectives* (pp. 93–116). Lanham, MD: Lexington.

UNESCO. (1997). *Adult education: The Hamburg declaration, the agenda for the future.* Fifth International Conference on Adult Education, 14–18 July.

UNESCO. (2005). *Guidelines for quality provision in cross-border higher education.* Paris: Author.

United Nations Development Program. (2001). *Human development report 2001: Making new technologies work for human development.* New York: Author.

wa Thiong'o, W.N. (2005). Europhone or African memory: The challenges of the pan Africanist intellectual in the era of globalization. In T. Mkandawire (Ed.), *African intellectuals: Rethinking political, language, gender and development* (pp. 155–164). Dakar: CODESRIA.

World Bank. (2001). *Constructing knowledge societies: New challenge for tertiary education, a World Bank strategy. Volume 2: Education Group, Human Development Network.* New York: Oxford University Press.

World Bank. (2003). *Poverty reduction strategy papers (PRSPs).* London: Bretton Woods Project.

Zeleza, P.T. (2004). Neo-liberalism and academic freedom. In P.T. Zeleza & A. Adebayo (Eds.), *African universities in the twenty-first century* (pp. 42–68). Dakar: CODESRIA.

CHAPTER 3

Cultural Perspectives in African Adult Education: Indigenous Ways of Knowing in Lifelong Learning

Ladislaus Semali

> The more I understand lifelong learning as a way of life, as a process of breaking barriers and of combating social exclusion, the more I am convinced that it is more about the rediscovery of Africa and the revival of traditional African pedagogy and values.
>
> (Avoseh, 2001, p. 479).

African adult education is closely related to the socioeconomic and political development of the countries where it is practiced. It is equally influenced by global trends in educational changes. Within the emerging global market economy and subsequent digital technology, schools are only one among many emerging educational agencies that prepare citizens for the world of work and lifelong learning. The lines between formal, informal, and nonformal education are blurring because of recent advances in access to information and technology. Both content and pedagogy are rapidly changing to meet the imperatives of Internet technologies and the emerging wireless environment that has collectively turned obsolete acclaimed teaching strategies and revolutionized the learning enterprise. People can now learn anytime and anywhere provided they have access to technology. Clearly, our beliefs about teaching and learning and how adults learn are changing as well.

In present times, even though many of us live in high-tech surroundings and make use of the most up-to-date information and communication techniques, the well-being of people is very much connected with the "wealth of nature" (Worster, 1994). Nature's wealth continues to inspire research, economic activities, and political action, but it also leads to conflict and even war. Nature's wealth is a place-based ecological resource that encompasses both the tangible and intangible assets of a community, including the wealth of knowledge resident there. Valuing nature's wealth is necessary for any adult education to take place (Semali, 1999), and, as intimated by Avoseh (2001) in the quote above, it is the first step toward innovation and discovery of wealth within a traditional African pedagogical context.

This chapter proposes a different approach to adult education that values nature's wealth in the African context. Rather than prescribing adult education as a set of fixed curricula to be learned outside the formal system, or following in the footsteps of North American educators who have defined adult education within a predominantly psychological framework, I suggest that we conceptualize African adult education broadly to encompass the reality of Africa. Instead of adult education as previously done in a separate presenting discipline of nonformal education, we are challenged to recognize the interrelationships among formal, nonformal, and informal education within a holistic program of lifelong learning. Instead of promoting adult education as lifelong learning based on formal schooling pedagogy, jobs, employment, and workplace productivity, I propose we talk about lifelong education as a way of knowing, thinking, and doing that enables indigenous innovation and discovery to build the capacities of farmers, farmers' groups, women's groups, youth, and communities to identify their needs, community assets, the collective capacity for innovation and creating new alternatives for resource-poor farmers, especially women. Building on farmer innovation, for example, implies a fundamental change in the roles of extension agents, health educators, and planners from transmitters of technical messages to facilitators of knowledge exchange between farmers and between farmers and extension officers themselves.

I discuss the epistemology surrounding adult education in these times of transition in sub-Saharan Africa. I raise questions about recent trends in globalization and what it means to indigenous local communities that occupy more than 80 percent of the African continent. How might a newly envisioned African adult education meet the needs of these populations, particularly to (1) build capacities of farmers' groups and rural communities in marginal areas to identify and evaluate market opportunities, (2) develop profitable enterprises, and (3) intensify production through experimentation while sustaining the resources upon which their livelihoods depend? As commonly practiced, adult education in

Africa has followed Western models of consumer trends and workplace productivity.

My task, therefore, will be first to discuss adult education at the time of transition. I argue that African countries shifted from their service-based policy agenda of the 1960s during the boom and bust period in the 1970s and 1980s, experienced the drastic effects of structural adjustments in the 1990s, and are now attempting to pursue an African renaissance agenda. It demonstrates how adult educators can help create deliberative democracy by working with civil society to engage African communities in public discourse and empower the citizenry as they struggle to reduce extreme hunger and poverty. Second, I outline ways to validate indigenous ways of knowing, thinking, and doing, and conclude by making recommendations for changes necessary to implement an indigenously informed African adult education.

African Adult Education in Transition

In the past few years, lifelong learning and adult literacy were highly valued by many African governments as the highest form of adult education. National literacy campaigns characterized the majority of the national development plans of African countries. Adult literacy was seen as key to national development, while an illiterate nation was seen as an underdeveloped society (Bhola, 1995; Semali, 1996) characterized by the conditionality of literacy in these terms: "Without literacy, development limps on one leg." The community's needs—including what was to be learned, when to learn it, and how—were determined by people from outside. In Tanzania, for example, government employees of the Institute of Adult Education, teachers, and extension agents made decisions of content and pedagogy. This approach was consistent with humanistic visions of lifelong learning at the time, such as those espoused by the United Nations Educational, Scientific, and Cultural Organization (UNESCO) in the late 1960s. In this era, education had deep social roots and was connected with democracy and self-development. Since the 1980s, however, there has been a gradual erosion of the commitment to equality and the total dominance of the economic imperative. The reasons for promoting adult education, for example, were now given only in economic terms, and nothing was said regarding issues of social justice or the process of enhancing feelings of self-efficacy in communities by identifying and removing conditions that reinforce powerlessness.

Many of these approaches tended to be top-down and lacked an effective process of community learning and empowerment. The decisions of what products and enterprises to develop or study and what markets to target were often prescribed by government agencies, private companies, or development organizations. These organizations then conducted a commodity market chain analysis and organized production to meet identified market demand and, often, external export market. These approaches have produced mixed results. While many studies have documented impressive results linking farmers to export markets, it has been argued that smallholder farmers have rarely benefited from these initiatives, as niche markets tend to be highly competitive and specialized, with rigorous quality standards that can be challenging to many small-scale farmers (Diao & Hazell, 2004). A new vision of African adult education can no longer be preoccupied with what people do not have or with disregard for what farmers know and have practiced for centuries.

The "Education For All" Movement

In the 1990s the Education for All (EFA) movement, first introduced at Jomtien in 1990 and ten years later revised by the Dakar Educational Goals, set goals for adult education. The third goal ensures that the learning needs of all young people and adults are met through equitable access to appropriate learning and life-skills programs, while goal four sets a target for achieving a 50 percent improvement in levels of adult literacy by 2015, with an emphasis on women, and equitable access to basic and continuing education for all adults. The underlying assumption of these goals is that the policy options outlined can bring long-term gains that will have a wider spin-off effect in moving toward quality education for all. An interim report issued in 2005 showed that there is an educational crisis looming large over sub-Saharan African countries for they lag behind other regions of the world in making progress toward the goal of providing education for all, particularly the target areas of universal primary education (UPE) and gender parity (UNESCO, 2005). It is predicted that the 23 African countries will not meet the 2015 EFA goals.

It is important to note that the increased reliance on EFA as a desirable goal for development is not new to sub-Saharan Africa. Since the 1960s, when the catchphrase was introduced during the anticolonial nationalist movements, African governments have singled out "education" as central to their rationale for national development. The assumption was that the benefits of education were easy to explain to the masses and that reforms in that field were likely to show results in a short time. Recently, for example, Kenyan politicians used access to UPE as a campaign promise in the 2002 elections for the current government. For them, education remains a high priority in the government's development strategy. Similarly, Tanzania saw a rapid increase in the net primary enrolment rate, from 57 percent to 85 percent in 2002, following the implementation of the campaign promise in the 2000 national election to abolish fees in primary schools (Kattan & Burnett, 2004).

What does the contemporary lifelong learning discourse mean to politicians and policy makers? We are reminded that, within the broader discourse, lifelong learning is only superficially expressive of the glories and campaign slogans or of the benefits to be reaped thereof. A Western view of lifelong learning permeates the African landscape. The progressive, ethical, and liberatory nature of lifelong learning is marginalized or excluded from the discourse, and it may best be seen, accordingly, as seriously regressive, counterethical, and nonliberatory. This view of Western lifelong learning is substantially lacking in critical concern, social vision, and any commitment to social justice and equity. It constructs education as a commodified private good, for which individuals should pay. It focuses strongly on individual interests and on vocational skills development. That education, which is funded by the state, is focused increasingly on the development of basic life skills and vocational skills in the interests of engagement in and service to the global economy. Educational engagement is increasingly seen as desirably embedded in the economically productive activities that are its desired outcomes, further limiting any opportunity for socially progressive learning. It is suggested that if the prevailing lifelong learning discourse is to be made more culturally progressive, in both its educational activities and its learning outcomes, it cannot be through a return to traditional progressive ideologies (Bagnall, 2000).

African Adult Education and the HIV/AIDS Epidemic

We cannot talk about education in Africa today without implicating the devastating pandemic. In the past few years, with the onset of HIV/AIDS and with poverty levels on the rise, governments are caught between tight budgets and a deadly disease. African governments' response to the pandemic has been mixed. The role played by adult education is mute. African governments are also caught in the swirl of sending mixed messages about the educational priorities for the masses, particularly in the public health sector and in HIV/AIDS prevention. Some critics suspect that such confusion is partly the result of a long-term legacy that seems to linger on forever. They contend that the interference of one worldview over another in the learning of science concepts is perhaps similar to the interference of a first language in the learning of a second (Jegede, 1999). However, a research study to investigate such interference in the cultural learning environments of diverse populations, including indigenous youth and adults, has not received much attention. Further, a project aspiring to integrate indigenous methods of teaching in the school curriculum, adult education programs, community health classes, and so on, is not popular among curriculum planners or adult educators. Collectively, these concerns present an educational crisis that spans the entire educational system including adult education.

This chapter attempts to situate adult education within a holistic view of knowledge (rather than in the previously segmented, or instrumentalist, as often departmentalized, into disciplinary silos.) In this view, knowledge is continuous and contextual. In this chapter, I envisage adult education to be constructed knowledge that has a complementary relationship with lifelong learning, informal education, and indigenous knowledge. This indigenous, or place-based, knowledge is part of a dynamic "cooperative innovation system" that continues to work—despite overwhelming pressures to destroy it—and continues to offer humankind an irreplaceable hope for planetary survival.

Indigenous Knowledge in the Context of Adult Education

Knowledge, understood as a set of shared, subjective, often taken-for-granted meanings and assumptions that are socially and historically constructed (Mezirow, 1995), is one of the hallmarks of human existence. Forbidden or otherwise, it has been central to our evolution as a species, as well as a key to assuming power. Belenky, Clinchy, Goldberger, and Tarule (1986), authors of *Women's Ways of Knowing*, remind us that our basic assumptions about the origins of knowledge shape the way we see the world and ourselves as participants in it. Perhaps it is well advised to recall here the groundbreaking works of Knowles (1990) on andragogy and self-directed learning, Mezirow's ideas of perspective transformation, and several other models that have contributed to our understanding of adult learning.

According to Knowles (1990), andragogy assumes that the point at which an individual achieves a self-concept of essential self-direction is the point at which he/she psychologically becomes an adult. Something very critical happens when this occurs: the individual develops a deep psychological need to be perceived by others as being self-directing. Thus, when he finds himself in a situation in which he is not allowed to be self-directing, he experiences a tension between that situation and his self-concept. His reaction is bound to be tainted with resentment and resistance. Knowles's assumptions (p. 57) include the following:

1. *The need to know*—adult learners need to know why they need to learn something before undertaking to learn it.
2. *Learner self-concept*—adults need to be responsible for their own decisions and to be treated as capable of self-direction.
3. *Role of learners' experience*—adult learners have a variety of experiences in life that represent the richest resource for learning. These experiences are however imbued with bias and presupposition.
4. *Readiness to learn*—adults are ready to learn those things they need to know in order to cope effectively with life situations.
5. *Orientation to learning*—adults are motivated to learn to the extent that they perceive it will help them perform tasks they confront in their life situations.

In sum, Knowles's formulation of the principles of andragogy may be taken as much as an integration or summation of other learning theories and therefore represents the assumptions and values underlying much modern adult educational theory. In addition, recent theoretical orientations, including consciousness and learning, situated cognition, critical theory, and feminist pedagogy, shed light to expand our understanding of indigenous pedagogies as the knowledge base of adult learning.

Over the last several centuries, a certain form of knowledge has dominated human existence. Sometimes labeled as instrumental knowledge, this Western, scientifically based knowledge has come to be seen as the most appropriate form of knowledge for a modern world. Other forms of knowledge, such as women's knowledge, indigenous knowledge, experiential knowledge, transformative knowledge, and local knowledge, have all been marginalized or excluded to a greater or lesser extent by the dominance of this instrumental knowledge, within both the academy and the society at large.

In his study of knowledge and human interests, Habermas (1978) challenged the dominance of instrumental forms of knowledge by presenting a framework that involved three knowledge domains. The first he called *empirical/analytic knowledge,* which Morrow and Torres (1995, p. 24) describe as being "based upon a desire potentially to control through the analysis of objective determinants." Instrumental knowledge fits into this category. How to calibrate a pesticide sprayer, or construct a survey of train employees, are all examples of empirical/analytic knowledge. Many of the African adult education programs of the 1960s and 1970s were conceptualized using this instrumental approach to learning and discovery. Having the ability to read instructional manuals and package labels were considered signs of being literate. While there was a turning point during the 1990s, the main feature of instrumental knowledge was that lifelong learning was less of a slogan and more of a tool for the reform and modernization of aspects of national education and training systems. Its rise has accompanied a wider transformation in the relationship between civil society and state in the Western nations (Field, 2001).

The second knowledge domain Habermas called the *historical-hermeneutic knowledge*. This is knowledge that Morrow and Torres (1995, p. 24) describe as "based upon a desire potentially to . . . understand through the interpretation of meanings." Discourse analysis, narratives, and "women's ways of knowing" are all forms of historical-hermeneutic knowledge. In the current debates that have dominated feminist perspectives, women's ways of knowing center and make problematic women's diverse situations and the institutions that frame those situations. Research topics include policy issues related to realizing social justice for women in special contexts or knowledge about oppressive situations for women (Olesen, 2000). In the African context, few examples have existed until very recently. However, building local capacity that values local knowledge and a community's ability to interpret their local situation, create their own local meanings, and set their own priorities is based on this second domain. Each society establishes a mode of existence, a distinct way of understanding itself, its activity, its history, and the ways to improve on the world it inhabits.

The third knowledge domain Habermas called *critical-emancipatory knowledge*. This involves knowledge "based upon a desire potentially to . . . transform reality through the demystification of falsifying forms of consciousness" (Morrow & Torres, 1995, p. 24). Transformative learning, critical reflection, and liberatory praxis are all part of critical-emancipatory knowledge. Emancipatory education was popularized in Africa after critical educators emerged, following the publication of radical adult educators, Paulo Freire (1971), Ivan Illich (1970), and others. On the one hand, Illich argues that for most people the right to learn is curtailed by the obligation to attend school. On the other hand, in *Pedagogy of the Oppressed*, Freire (1971) argues against the *banking concept of education* in favor of a liberatory, *dialogical* pedagogy designed to raise individuals' consciousness of oppression and to in turn transform oppressive social structures through "praxis." Freire was instrumental in the planning of several adult literacy campaigns in Tanzania, Ethiopia, and Cape Verde, and the influence of his writings lingers on throughout the continent still today.

The principles guiding my reflection and thinking in this chapter derive from notions of transformative pedagogy along Freire's notions of "unity in diversity." The assumption is that, as adult learners read the word and the world critically through multiple lenses, they can decolonize knowledge production from the hierarchies and dynamics of power (Freire & Macedo, 1987). In contrast to the predominantly single focus on the first knowledge domain, Habermas (1978) argues that humans need all three kinds of knowledge. This tripartite epistemology opens the way for understanding adult education, not only as an instrumental activity, but also as a deeply experiential form of meaning making and as an emancipatory practice that can lead to sustainable ways of life.

Globalization, Adult Education, and Valuing Indigenous Ways of Knowing

Globalization brings other issues to the fore of lifelong learning. It threatens to appropriate the shared, collective knowledge of non-Western systems into the private, proprietary knowledge of a few. Globalization cannot, however, function in a moral vacuum. The goals of education for cooperation and sustainable human development need to be clarified; new social contracts that can bind together democratic citizenship, social justice, and capitalism need to be developed or strengthened. Communities need to be stimulated in a manner that builds on what they have—including the knowledge, skills, and competencies they have acquired over centuries through indigenous methods.

Globalization is challenging the relevance, appropriateness, and effectiveness of existing modes of research, teaching, and university-community partnership building. The mighty force of globalization transcends natural boundaries, resulting in an increasingly borderless flow of goods and services, money, skilled and qualified manpower, information, and culture. This force threatens the survival of local species, cultures, and indigenous ways of knowing and doing (Shiva, 1993). The process of globalization also threatens place-based knowledge when local or indigenous knowledge is not employed as a complement to science-based knowledge generated by those outside a local community. Engaged universities will need to become proficient in developing support systems for community-based learning that complement more traditional approaches to knowledge transfer. The challenge is to coordinate these two ways of knowing in an iterative, integrative fashion that not only enables practice to proceed from theory but also enables theory to be generated from local practice (Hassel, 2005).

It is within this framework that rural communities and adults everywhere should be engaged. The preservation of inherent dignity in indigenous communities, enhancing their sense of self-respect, and in turn respecting their autonomy of choice and action—even when this means their rejection of a particular mode of education—must be given priority. Each society furnishes its own "construction" of the world, creating its own world in the sense that it invests "what is" with its distinctive meaning. Each society establishes a mode of existence, a distinct repertoire of meanings and its own history. Adult education and lifelong learning should service these aspirations.

For example, results of action research applying the Enabling Rural Innovation (ERI) approach in pilot sites in Malawi, Uganda, and Tanzania show that small-scale farmers know a lot about their environment; they are not ignorant or illiterate as previously assumed when it comes to knowledge about their local history, information about flora and fauna, and application of local medicines to humans and animals to cure diseases endemic to the community. They make rational decisions about their lives. They are not always attracted by higher economic returns. Rather, they use a range of economic and noneconomic criteria for selecting existing crops and livestock for new markets, as well as new crops for new markets. Evaluation of market opportunities stimulates farmers' experimentation to reduce risks, access new technologies, and improve the productivity and competitiveness of the selected enterprises (see Sanginga et al., 2004).

At present, the West's primary domination of the world lies in its monopolization of the very terms by which value is conceived, and its domination of the basic institutions that codify social life. The deculturation of the dominated societies is shown by the fact that they increasingly voice their predicaments and aspirations solely in terms of the categories sanctioned by the invading culture. This entails, at the limit, the asphyxiation of the recipient culture, and the loss of vitality and coherence of indigenous cultural forms. African societies are, under these conditions, made to feel that they have given little or nothing to others, as the contributions they have made to global knowledge continue to be ignored.

However, with the growing recognition of the value and importance of indigenous and place-based knowledge as cultural capital (Bourdieu & Passeron, 1986; Monkman, Monald, & Théramène, 2005) for sustainable development (Irwin, 1995) and as a basis for "prior knowledge" (Fordham, 1996; Shapiro, 2004) or schemas in learning (Bartlett, 1932), both the number of projects and the amount of information on indigenous knowledge have increased (see Dei, Hall, & Rosenberg, 2000; Semali & Kincheloe, 1999; Shiva, 1993; Warren, Slikkerveer, & Brokensha, 1995; World Bank, 2004;). Despite all these efforts, development projects, university curricula, and nonformal educational programs still appear to make little use of this valuable resource. But the situation is slowly changing as universities have come to accept some level of responsibility for bridging the gap between community-based and campus-based knowledge systems.

The Future of African Adult Education

What might the future of African adult education look like? How different must it be from current practices to make a difference? For the most part, the target population of African adult education programs will be adults living in rural areas, many of whom are engaged in subsistence farming. Their home environment continues to be impoverished, and lacks permanent housing, safe water, and food security. With these conditions affecting many Africans, it becomes inconceivable how adults living under such circumstances can have the motivation to remain literate or pursue independent learning. What might lifelong learning mean for them? According to Nyerere (1978, p. 28), "the main purpose of adult education is to help people develop themselves, and enable them to examine the possible alternative course of action; to make a choice between those alternatives in keeping with their own purposes, and it must equip them with the ability to translate their decisions to reality."

Clearly, Nyerere's view on the purpose and role of adult education in development is akin to Freire's analysis of education: education is either for liberation or for domestication. Nyerere's analysis of the negative effects of colonial education is also similar to Freire's critique of banking education and pedagogy of the oppressor or colonizer. Both Nyerere and Freire emphasize the raising of people's consciousness as the critical function of education for liberation. Therefore, it seems to me that the future of African adult education cannot be outside the quest for liberation from hunger, disease, and HIV/AIDS. African adult education should continue to promote change at the same time as it assists people in controlling both the change that they induce and "that which is forced upon them by the decisions of other men or the cataclysms of nature" (Nyerere, 1978, p. 29). The lessons learned from Nyerere's *Education for*

Self-Reliance (ESR) in Tanzania are illustrative of the difficulties of indigenizing education, and therefore the future of African adult education will be neither easy nor a panacea.

Though not completely successful, the ESR experiment aimed to localize the curriculum in Tanzania by emphasizing practical rural-oriented education. The program of ESR was more ideological than pedagogical, however. The political view at the time of its formulation was that Western forms of education had caused harm to African traditional ways of learning and teaching, and therefore needed to be de-emphasized. In particular, the civics syllabus for secondary education until today emphasizes that the aims and objectives of secondary education should include indigenous knowledge—to promote the acquisition and appreciation of the culture, customs, and traditions of the people of Tanzania, and to enhance further development and appreciation of national unity, identity and ethics, personal integrity, respect for human rights, cultural and moral values, customs, traditions, and civic responsibilities and obligations and readiness to do both mental and manual work.

Nyerere's (1968) radical ideals about indigenizing education may have simply echoed at the time the political climate of educational reform that was taking place elsewhere in Africa, particularly in the West African countries of Nigeria and Ghana. Elsewhere, though much later in the postcolonial era, conferences, study seminars, and high-level government discussions about Africanization and the reintroduction of indigenous education in formal schooling were undertaken in Botswana, Kenya, Guinea, Uganda, Zaire, Zambia, and Zimbabwe, among others, but nowhere has the "traditional" African education component become as apparent in policy documents as it was in ESR (Fafunwa & Aisiku, 1982).

Decolonization: (Re)Valuing African Peoples and Their Heritage

The future must include the process of decolonization of the mind (Wa Thiong'o, 1986). One place to start the decolonization of knowledge and building for the future is for administrators and government agents to encourage adult education curriculum developers and extension agents to rethink adult education and begin a new path that departs from foreign imitations and interpretations of what is important at the local level.

As discussed in this chapter, we can identify three themes that illustrate how the future of an indigenized adult education will look like. The first theme has to do with valuing and (re)valuing African heritage. Nyerere's writings emphasized the role of adult education in arousing awareness and consciousness among the people and the need and possibility for change. The resilience of the African people can be attributed to the wealth of their heritage—cultural, geographical, natural resources, and so on. Valuing and revaluing these tangible and intangible resources must be the source of future development. Continuing to depend on outside help and the interpretation of the world or recounting of African history, as it is currently done in books, mass media, and school curricula, will not leave room for the possibility for the kind of change envisioned by Nyerere.

To begin indigenizing African adult education, both educators and learners have to reach out to African history and traditions and rediscover African values and indigenous epistemologies. Unfortunately, those of us who work in adult education have condoned behavior and attitudes that continue to despise and devalue heritage studies. We have also been extremely unkind to people who are not literate, including rural populations, pastoralists, hunters and gatherers, and the disabled. Several publications in our purview, whether we know it or not, have published derogatory anecdotal information that carry colonial labels that are demeaning—for example, labels like Pygmy peoples of Rwanda, instead of the Batwa; or Bushmen, instead of the San of South Africa and Namibia—or devalue the lifestyles of pastoralists such as the Maasai of Kenya and Tanzania, the Mursi of Ethiopia, and the Tuareg of West Africa, whose ways of life require the ability to move freely among traditional pasture lands to graze their animals. These groups are less characterized by their mode of production than by their small numbers, remote locale, lack of representation in political structures, and the extreme threat to their lands and lifestyles from governmental and international interests that have failed to consult them in decisions concerning their fate.

Furthermore, we have done little to improve the low self-perception among these populations as learners. They have been classified as "illiterate." For them, illiteracy is an affliction and encumbrance. It has been targeted for elimination in campaigns based on overtly militaristic or pathological paradigms that seek to "eradicate," "attack," "wipe it out," and so on. We cannot ignore how we still equate illiteracy with absolute ignorance, viewing what is not written as thoughtless and, at its limit, as primitivism. As a result, some of us have engaged in strategic disempowerment even when we have intended to empower (Hoppers, 2002). It does not hurt—except our egos, of course—to acknowledge that people or groups have a heritage of knowledge that is the basis of their interpretation of the world, which they use in specific situations and in accordance with procedures they recognize. Education and training of various types should begin here, in people's lives, among their peers, with their style, their customs, their decisions, and their utopian ideals. Any African adult educational enterprise must begin by discovering and recognizing this foundational principle. For this reason, as a mediator of knowledge, the adult educator should focus not on the transfer of knowledge but rather on fostering an epistemological relationship between forms of knowledge and the subjects in the knowing process. It is by fostering this essential link that grassroots and indigenous groups can locate themselves and their knowledge within the larger context of knowledge and power relations.

The Troubling Scourge of a Disease: HIV/AIDS

The second theme and message is that there is no future for the people of Africa without finding prevention and cure to the devastating pandemic. African adult education must target this health hazard. Experts agree that prevention through education is the best way to fight the transmission of HIV and that education must begin before young people initiate sexual activity and certainly no later than seventh grade (Hartell, 2005). Unless we understand how teachers construct knowledge, we can hardly claim to understand teaching or the pedagogical constructs that form much of the teaching that takes place in formal school or in adult education classrooms. The ways in which schools and their stakeholders produce and reproduce knowledge about HIV/AIDS must be confronted.

The most important lesson to take from this chapter, however, is that teacher knowledge continues to be contradicted by African epistemologies, local customs, taboos, and traditional beliefs and practices. Epidemiologists and educators must bridge this gap in their educational materials. Regardless of whether an adult educational program imparts a given set of life skills, if an intervention fails to produce among learners an accurate knowledge of the basic characteristics of HIV and its prevention, researchers are mistaken to expect that it will lead to significant positive behavior change. We are reminded that the extended family continues to be the major resource that most victims and orphaned children rely on for knowledge, emotional support, and material resources. The extended family has to be the focus and target of adult education programs.

Creativity, Discovery, and Innovation

The third and final theme is the recognition that African lifelong learning is characterized by creativity, discovery, and innovation. Any attempt to reinvigorate adult education in local communities to reduce poverty and hunger or to prevent HIV/AIDS must value and recognize local resilience that is fuelled by creativity, discovery, and innovation. For example, linking farmers to growth markets is an important strategy for improving the adoption of agricultural technologies as well as for raising rural incomes and reducing poverty. However, until recently, one critical gap in agricultural research and development has been its failure to link farmers to profitable markets and to increase incomes for marketing agricultural products. National governments in Uganda, Malawi, and Tanzania are increasingly placing emphasis on transforming subsistence agriculture by taking farming as business and by infusing an entrepreneurial culture in rural communities in diverse situations, which can help achieve their income and other livelihood aspirations through better links with markets. However, what is not so obvious in this effort is how to link small-scale farmers in marginal areas to expanding markets, and how to develop methods and approaches that effectively integrate research and marketing and enterprise development. This approach ought to be the focus for the future of African adult education.

To conclude this chapter, I quote from *Learning: The Treasure Within,* in the words of Jacques Delors, (1996), who suggested the four pillars of learning: learning to know, learning to do, learning to live together, and learning to be. These four pillars represent the African metaphor of the four-legged stool. The African people's quest for lifelong learning is immersed in the nature's wealth of indigenous knowledge and African traditions, where knowing, thinking, being, doing, and living together in the extended family is the hallmark of African identity.

References

Avoseh, M.B. (2001). Learning to be active citizens: Lessons of traditional African for lifelong learning. *International Journal of Lifelong Education, 20*(6), 479–486.

Bagnall, R. (2000). Lifelong learning and the limitations of economic determinism. *International Journal of Lifelong Education, 9*(1), 20–35.

Bartlett, R. (1932). *Remembering.* Cambridge, MA: Harvard University Press.

Belenky, M.F., Clinchy, B.M., Goldberger, N.R., & Tarule, J.M. (1986). *Women's ways of knowing: The development of self, voice, and mind.* New York: Basic Books.

Bhola, H.S. (1995). Without literacy, development limps on one leg. *Adult Education and Development, 24*(3), 68–71.

Bourdieu, P., & Passeron, J. (1986). The forms of capital. In J.G. Richardson (Ed.), *Handbook for theory and research for the sociology of education* (pp. 241–258). Westport, CT: Greenwood.

Dei, G., Hall, B., & Rosenberg, D.G. (2000). *Indigenous knowledge in global contexts: Multiple readings of our world.* Toronto, ON: University of Toronto Press.
Delors, J. (1996). *Learning: The treasure within.* Paris: UNESCO.
Diao, X., & Hazell, P. (2004, April 3). Exploring market opportunities for African smallholders. Paper prepared for the 2020 Africa Conference, *Assuring food security in Africa by 2020: Prioritizing actions, strengthening actors, and facilitating partnerships.* Kampala, Uganda.
Fafunwa, A.B., & Aisiku, J.U. (1982). *Education in Africa.* Boston: George Allen & Unwin.
Field, J. (2001). Lifelong education. *International Journal of Lifelong Education, 20*(1–2), 3–15.
Freire, P. (1971). *Pedagogy of the oppressed.* New York: Herder and Herder.
Freire, P., & Macedo, D. (1987). *Literacy, reading the world and the word.* Amherst, MA: Bergin & Garvey.
Habermas, J. (1978). *Knowledge and human interests* (J.J. Shapiro, Trans.). Boston: Beacon Press.
Hartell, G. (2005). HIV/AIDS in South Africa: A review of sexual behavior among adolescents. *Adolescence, 40*(157), 171–181.
Hassel, G. (2005). The craft of cross-cultural engagement. *Journal of Extension, 42*(6). Retrieved July 14, 2006, from http://www.joe.org/joe/2005december/index.shtml.
Hoppers, C.O. (2002). Indigenous knowledge systems: The missing link in literacy, poverty alleviation and development strategies in Africa. *Africa Insight, 32*(1), 3–7.
Illich, I. (1970). *Deschooling society.* New York: Harper & Row.
Irwin, A. (1995). *Citizen science: A study of people, expertise, and sustainable development (environment and society).* London: Routledge.
Jegede, O. (1999). Science education in non-Western cultures: Towards a theory of collateral learning. In L. Semali & J. Kincheloe (Eds.), *What is indigenous knowledge? Voices from the academy* (pp. 119–142). New York: Garland.
Kattan, R., & Burnett, N. (2004). *User fees in primary education.* Washington, DC: World Bank.
Knowles, M.S. (1990). *The adult learner: A neglected species* (4th ed.). Houston: Gulf Publishing.
Mezirow, J. (1995). Transformation theory of adult learning. In M.R. Welton (Ed.), *In defense of the lifeworld: Critical perspectives on adult learning* (pp. 39–70). Albany: State University of New York Press.
Monkman, K., Monald, M., & Théramène, F.D. (2005). Social and cultural capital in an urban Latino school community. *Urban Education, 40,* 4–33.
Morrow, R.A., & Torres, C.A. (1995). *Social theory and education: A critique of theories of social and cultural reproduction.* Albany: State University of New York Press.
Nyerere, J. (1968). Education for self-reliance. In *Freedom and socialism/Uhuru na Ujamaa: Essays on socialism* (pp. 278–290). New York: Oxford University Press.
Nyerere, J. (1978). Development is for man, by man and of man: The declaration of Dar-es-Salaam. In B. Hall & J.R. Kidd (Eds.), *Adult learning: A design for action* (pp. 27–36). Oxford: Pergamon Press.

Olesen, V.L. (2000). Feminism and qualitative research at and into the millennium. In N.K. Denzin & Y.S. Lincoln (Eds.), *Handbook of qualitative research* (2nd ed., pp. 215–255). Thousand Oaks, CA: Sage.

Sanginga, P.C., Best, R., Chitsike, C., Delve, R., Kaaria, S., & Kirby, R. (2004). Enabling rural innovation in Africa: An approach for integrating farmer participatory research and market orientation for building the assets of rural poor. *Uganda Journal of Agricultural Sciences, 9*(1), 942–957.

Semali, L. (1996). *Postliteracy in the age of democracy. A comparative study of China and Tanzania.* San Francisco: Austin & Winfield.

Semali, L. (1999). Community as classroom: Dilemmas of valuing African indigenous literacy in education. *International Review of Education, 45*(3–4), 305–319.

Semali, L., and Kincheloe, J. (1999). *What is indigenous knowledge? Voices from the academy.* New York: Garland.

Shapiro, A.M. (2004). How including prior knowledge as a subject variable may change outcomes of learning research. *American Educational Research Journal, 41*(1), 159–189.

Shiva, V. (1993). *Monocultures of the mind: Perspectives in biodiversity and biotechnology.* London: Zed Books.

United Nations. (2005). *The millennium development goals report, 2005.* New York: Author.

United Nations Educational, Scientific, and Cultural Organization. (2005, October 7–8). *Ministerial Round Table, Education for All.* Paris: Author.

Warren, M.D., Slikkerveer, L., & Brokensha, D. (1995). *The cultural dimension of development: Indigenous knowledge systems.* London: Intermediate Technology Publications.

Wa Thiong'o, N. (1986). *Decolonizing the mind: The politics of language in African literature.* New York: Heinemann.

World Bank. (2004). Indigenous knowledge: Pathways to global development. *Marking five years of the World Bank Indigenous Knowledge for Development program.* Washington, DC: IK Notes—Knowledge and Learning Group.

Worster, D. (1994). *The wealth of nature: Environmental history and the ecological imagination.* New York: Oxford University Press.

CHAPTER 4

Mwalimu's Mission: Julius Nyerere as (Adult) Educator and Philosopher of Community Development

Christine Mhina and Ali A. Abdi

Julius Kambarage Nyerere, Tanzania's late president and philosopher-statesman who passed away in 1999, was born on April 13, 1922, in Butiama, a village in Northern Tanzania. As a young boy growing up in the open plains of East Africa, Nyerere and his peers, in both occupational and socioenvironmental categories, absorbed unstructured clusters of informal education from family, relatives, and other educated members of their society, in addition to certain age-based rituals that were common to their cultural and political situation. After he attended the Tabora government school, he went on to Makerere University in Uganda, and eventually received a Master's of Arts degree from the University of Edinburgh in Scotland. The simple points about Nyerere's early upbringing are important in regard to one primary (perhaps unanswerable) question: What would have Nyerere's life looked like had colonialism never came to Tanzania? Indeed, a very hypothetical question, but one that remains relevant to his long struggle for Tanzania's freedom, educational attainment, and overall social development.

Undoubtedly the influence of Nyerere on the both the theoretical and practical sides of his country's education is immense. As far as we know, no other leader in the African continent, or perhaps elsewhere, has preoccupied him or herself with issues of educational problems and prospects in the last 40 years or so more than the man Tanzanians and the rest of the world came to know as *Mwalimu* (teacher), a title that itself speaks volumes about the late leader's commitment to what he termed "education for self-reliance" (Nyerere, 1968). More than that, Nyerere was a creative educational thinker, a man who studied situations and tried, as much as possible, to create an educational vision and plan that was, even by hindsight, rich with pragmatic possibilities. His educational philosophy was premised on the fundamental need to liberate people from the restraints and limitations of ignorance and dependency. As Smith (1998) noted, Nyerere's educational philosophy can be classified under two main headings: general education for self-reliance on the one hand, and specialized adult education on the other. In this chapter, we engage as a general perspective, discussions and analyses of Nyerere's proposals for education and social development; as should be expected, we will also entertain some of his detractor's points and criticisms. While the generic term *education* will be used, it should be understood that in all his educational designs, Nyerere was adamant on including the learning of adults in education programs so all people could become enlightened, productive members of their societies.

When Tanganyika became an independent country in December 1961, Julius Nyerere was sworn in as the country's first postcolonial president. By the time he became the first president of United Republic of Tanzania in 1964, he already knew the damage that colonialism had brought to his country and to Africa: the destruction of local development schemes and the deliberate devaluing of indigenous knowledge systems which formed the basis of the continent's traditional educational schemes. Immediately, then, Nyerere was clear on the expansive inappropriateness of the formal learning systems that were inherited from colonialism for peoples who had different historical and general livelihood arrangements. Instead of serving the purpose of development, Nyerere (1968) viewed the inherited education system as an instrument of underdevelopment and as elitist in nature, mainly catering to the needs and interests of a minority of its clients with the social status, income, and privileges that accompanied employment in elite occupations. Furthermore, he realized, the colonial education system divorced its participants from the society for which they were supposed to be trained.

Nyerere's greatest challenge was to come up with basic educational philosophies, policies, and projects that would be relevant to a poor, developing country aspiring for a self-reliant, socialist mode of development. The pertinence of socialism need not detain our thinking more than is warranted; while ideological adherence to the program of Marxism is not and should not be intended here, it is also the case that Nyerere and others (e.g., Ghana's Nkrumah and Senegal's Senghor) have described, with different attachments and emphases, an African socialism that mainly spoke about a historical prerogative where people were, by-and-large, communal in the management and distribution of their resources (Abdi, 2002). As Nyerere wrote, "the objective of socialism in the United Republic of Tanzania is to build a society in which all members have equal rights and equal opportunities; in which all can live with their neighbors without suffering or imposing injustice; being exploited or exploiting" (1968, p. 340).

Nyerere set out some of the most important elements of his vision for education in the oft-referenced "Education for Self-Reliance" essay which became a part of his well-known book *Freedom and Socialism* (1968), in which he articulated his strategic plans for reducing dependence by promoting self-reliance. He believed that education should not be considered a special privilege that was to be found only in the preserves of the elite or even primarily an investment in human capital, but a fundamental right of all citizens (Nyerere, 1968; Samoff, 1990). Further, education had to address the realities of life as it was to be lived in Tanzania. In his essay "Education Never Ends," Nyerere (1979) wrote:

> This is what our educational system has to encourage. It has to foster the social goals of living together, and working together for the common good. It has to prepare our young people to play a dynamic and constructive part in the development of a society in which all members share fairly in the good or bad fortune of the group, and in which progress is measured in terms of human well-being, not prestige buildings, cars or other such things, whether privately or publicly owned. Our education must therefore inculcate a sense of commitment to the total community, and help the pupils to accept the values appropriate to our kind of future, not those appropriate to our colonial past. (pp. 20–21)

In Nyerere's thinking, education was a tool to serve the common good, that would simultaneously foster cooperation and promote equality. While many, especially those who tow the line of global capitalism where ideology takes precedence over pragmatic analysis, may dismiss Nyerere's fundamental objectives here as a grandiose socialist dream that missed the realistic outcomes of education, others would beg to disagree (see Abdi, in press; Rahnema, 1976 Samoff, 1990;). Samoff summarizes the Tanzanian educational reform, which contextually speaking should be considered a success, by saying:

> In a brief period, Tanzania, regardless of being poor, introduced institutional changes that reached nearly all its citizens. Primary education in

> Tanzania is essentially universal. Initial instruction uses a language and draws on experiences and materials that are familiar to everyone. The National board has a responsibility of setting and marking examinations. Tanzania adult literacy is now among the highest in Africa. (p. 209)

Specifically addressing adult education, Nyerere and his cohorts saw mass literacy as the sine qua non for effective citizen participation in both the policy and productive segments of society. As Nyerere repeatedly stressed, Tanzania's development could not await the education of a new generation; education had to focus on adults as well as children. He called for adult education to be directed at helping people to help themselves (as implicated in the theory and practice of self-reliance) and to be approached as a part of life (Smith, 1998). It was due to Nyerere's unflinching prioritization, therefore, that adult education in Tanzania was given a very special space within the country's overall development plans. Literacy programs were community-based and had substantial grassroots involvement. In the late 1960s and into the early 1970s, mass literacy campaigns were initiated throughout the country; for example, the Man Is Health campaign in 1973 and the Food Is Life campaign in 1975, while Folk Development Colleges (FDCs) were created to provide postliteracy programs and learning centers (Samoff, 1990). Between 1975 and 1977, illiteracy rates fell dramatically, and reached a low point of 9.6 percent in 1986 (Smith, 1998).

Nyerere's philosophy of adult education resonated with the concepts of conscientization later popularized by Paulo Freire as empowerment and liberation. Nyerere argued that "the first function of adult education is to inspire both a desire to change, and an understanding that change is possible" (1976, p. 29). In his speeches, Nyerere emphasized the significance of adult education in raising consciousness and critical awareness among the people about the need to combat what he perceived as a "sense of fatalism" among Tanzanian people (Mayo, 2001). Nyerere emphasized that being poor and backward should not be perceived as permanent God-given conditions since such situations are, without exception, transformable. The aim of adult education in this context was "to liberate the mind in a way that allows people to develop a strong sense of agency characterized by a belief in their ability to master circumstances" (Mayo, 2001, p. 199). It should be emphasized that building a strong sense of agency entails the involvement of the learners in their own education.

Thus, Nyerere's view of adult education stretched far beyond the classroom. His adult educational philosophy was learner-centered, emphasizing that adult learners have to participate in identifying their own learning needs and interests, and that their learning needs should be centered in their own problems and experiences. Long before the now popularized foci on the dialogic method, Nyerere criticized the type of education where the teacher is seen as the sole transmitter of knowledge by saying "the teacher of adults is not giving to another something that he possesses, . . . instead the teacher helps the learner to develop his [her] own potential and his [her] own capacity" (1976, p. 30). Adult learners' potential and capacity include their needs and the knowledge they have, which Nyerere considered as the fundamental determinant of the adult education method. Nyerere (1979, p. 53) stated that "every adult knows something about the subject he is interested in, even if he is not aware he knows it," an idea which is in tune with literature on tacit knowing (Polanyi, 1968).

Literature supporting the idea that local people have vast knowledge of their situation is abundant. For instance, Freire (1985, p. 14) declares: "Human ignorance and knowledge are not absolute. No one knows everything; no one is ignorant of everything." Futhermore, Polanyi (1962, 1966) believes that people know more than they can tell, since we engage in tacit knowing through virtually anything we do. Likewise, some theorists in the discipline of adult education support the idea of adult learners possessing a lot of knowledge. One of the assumptions of andragogy is that adults bring with them a depth and breadth of experience that can be used as a resource for theirs and others' learning (Merriam & Cafarella, 1999). Mezirow (1990) believes that the subject matter of transformative adult education is the learner's experience; that is, knowledge is acquired through informal learning.

Coffield (2000) and Polanyi (1968) have documented the importance of informal learning through social networks and other learning opportunities outside the context of formalized adult learning. They argue that, as much as informal education is unorganized and often unsystematic, it still accounts for the great bulk of any person's total lifetime learning. It is in this context that Nyerere emphasized listening to and building on the knowledge that adult learners possess. Drawing and building on things learners already know, combined with what adult educators know, should result in what Nyerere called "mutuality of learning"—a sharing of knowledge that extends the totality of our understanding of issues pertaining to our lives. Only on this basis of sharing and equality, Nyerere (1979) stressed, "[is it] possible to make full use of the existing human resources in the development of a community, a village, or a nation" (p. 53).

Linking Values to Pragmatism

One example of the power of informal education is how Nyerere's upbringing affected his life and worldview. In many of his speeches, including interviews where he had to respond to difficult questions spontaneously, the integration of rich informal knowledge was brought to surface. For instance, Nyerere drew insight from his experiences as a family member to construct a theory on women's liberation while writing an essay on "The freedom of Women" in 1974. Nyerere admitted that the British philosopher John Stuart Mill, who had written about the subjugation of women in the nineteenth century, had influenced his understanding and views of the issue, but contextually added that as his father had many wives, he knew how hard they had to work and what they went through as women (interview by Ikaweba Bunting, 1998). Nyerere, without pretence, empathized with their situation. Nyerere has also been influenced by John Stuart Mill's theory on women's subjugation, which triggered his latent knowledge (Polanyi, 1968), which he had previously acquired informally (Coffield, 2000) (see Figure 4.1).

The combination of Nyerere's personality of caring for people with his informally acquired knowledge about women's subjugation was later conceptualized in his theories and writing not only on women's freedom in Tanzania, but also in regard to the liberation of the African people, many of whom were still under direct colonial control. As a way of prag-

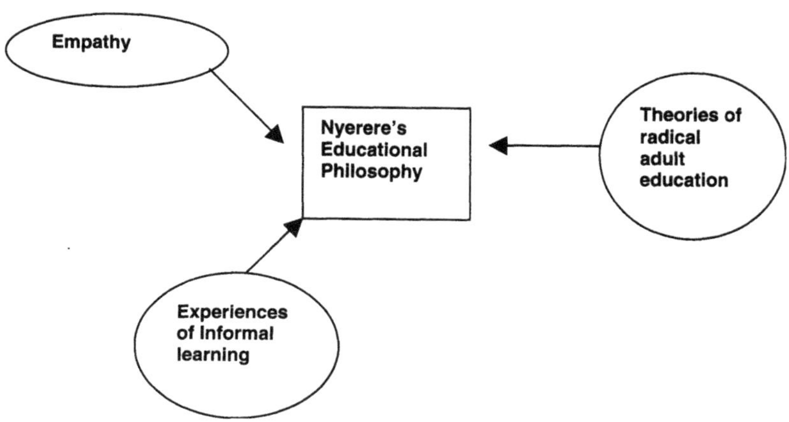

Figure 4.1 Values informally learned combined with academic theories

matically philosophizing the connectivity of societal needs and intersections, the subjugated situation of women represented for Nyerere, tangible disturbances in the corporeal oppression of the community being either under colonial rule or via other regimes of deprivation and marginalization. Moreover, Nyerere's tendency and energy to seek solutions that dignify humanity may also be attributed to the outcome of his informal learning and the long-term process of awareness-raising that he went through during his lifetime. During his interview with Bunting, Nyerere reported that the idea of liberation was a process—something that developed with time and grew inside him gradually. He said,

> What many of us went through was simply a desire to be accepted by the white man. . . . A kind of inert dissatisfaction that we were not accepted as equals. . . . But it was events in Ghana in 1949 that fundamentally changed my attitude. When Kwame Nkrumah was released from prison this produced a transformation. I was in Britain and oh you could see it in the Ghanaians! They became different human beings, different from all the rest of us! This thing of freedom began growing inside all of us. (Bunting, 1998)

In this dialogue, Nyerere demonstrates how he went through a process of conscientization, moving toward the idea of freedom theoretically, emphasizing something that exists "in the mindset of improving the lives and welfare of Africans" (Bunting, 1998).

Nyerere was a man who had deep religious convictions. He successfully translated these convictions and complementary personality traits into demonstrable concerns about freedom, justice and equality, the alleviation of hunger, poverty, and disease. Throughout his leadership, Nyerere worked hard to seek solutions for sustainable development and tried to overcome the malaise of colonialism and postcolonial modernity, the latter remaining an exclusionist project in human advancement terms. Unlike his experiences under colonial rule, Nyerere emphasized that Africans in particular have to maintain their struggle for justice, equality, and freedom within the independent countries of the continent. He offered sanctuary in Tanzania to members of African liberation movements from South Africa, Zimbabwe, Mozambique, Angola, and Uganda, and in 1978, he sent Tanzanian troops to depose the destructive regime of Ugandan dictator Idi Amin. In addition, Tanzania hosted the African Liberation Committee from its inception in 1963. It was Nyerere's celebrated role and efforts in African decolonization and his overall human rights focus that the African National Congress' (ANC) (South Africa) honored him on his death, in a statement that partially read: "a legacy in his own lifetime, he served as a symbol of inspiration for all African nations in their liberation struggles to free themselves from the shackles of oppression and colonialism" (ANC, 1999). As former American president Jimmy Carter noted, the roots of Nyerere's attachment to equality can be seen, as we indicated earlier, as emanating from a combination of his religiosity and cultural background, horizontally colored by his conscientious attachment to the traditional heritage of most Tanzanians, which were translated, in his mind and actions, to his aversion to racism and colonialism (Pratt, 1999).

In addition to the above-mentioned attributes, Nyerere shared a preoccupation with economic development with almost all of the Third World leaders of his generation. From Nehru in India to Nkrumah in Ghana and Manley in Jamaica, postcolonial leaders of colonized populations attempted to fulfill their people's wishes to fully enjoy the improvements in personal and social welfare brought about by economic development (Pratt, 1999). Generally, Nyerere's development policies were based on the goals of egalitarianism and human-centered development. Pratt summarizes Nyerere's central domestic preoccupation as follows: (1) developing the Tanzanian economy; (2) securing and retaining national control of the direction of Tanzania's economic development; (3) creating political institutions that would be widely participatory and sustain the extraordinary sense of common purpose; and (4) building a just society that would be genuinely equitable for all its citizens. The socialism he believed in was people-centered in the full sense of the phrase. It was characterized by personal and regional economic self-reliance, complemented by localized rural development projects. Nyerere strongly believed and stressed that development was about all people and not just a small and privileged minority. Pratt notes that the equality that Nyerere valued was different from the Western perspective of equality, which emphasizes initial opportunities for autonomous individuals. Rather, it was a holistic theory of equality to be enjoyed by all in a closely integrated and caring society where no one falls through the proverbial but practically painful tracks (or cracks).

Nyerere's concept of grassroots democracy, on the other hand, which revolved around Ujamaa, entailed a process of participatory democracy (Mayo, 2001). He was immensely attentive to the needs of the people at the bottom of the heap and had tremendous faith in the capacity of rural African peoples and their traditional values and ways of life. Nyerere's vision of development was set out in the Arusha Declaration of 1967 (reprinted in Nyerere, 1968):

> The objective of socialism in the United Republic of Tanzania is to build a society in which all members have equal rights and equal opportunities; in which all can live in peace with their neighbors without suffering or imposing injustice, being exploited, or exploiting; and in which all have a gradually increasing basic level of material welfare before any individual lives in luxury. (p. 340)

In his speeches, Nyerere stressed that Ujamaa villages were intended to be socialist organizations created by the people, and governed by those who lived and worked in them. They were not to be created or governed from outside. No one could be forced into an Ujamaa village, and no official at any level could go and tell the members of an Ujamaa village what they should do together (Nyerere, 1968). As much as Nyerere's social, political, and economic values display his commitment to bring change to Tanzanians and Africans in general, his policies were not as successful as was expected. In the next section we analyze what might have gone wrong.

Observed Gaps between Nyerere's Policies and Program Implementation

Although Nyerere is fondly remembered as a benign humanist and visionary, in his latter years of leadership it became clear that not many people understood his ideas well enough to implement them. Problems with the important implementation phase of the former president's programs might be attributed partly to inadequate human and material resources, but were also largely due to low-level understanding of the policies themselves, complemented by opposing forces that were always bent on thwarting the policies. Another mitigating factor was the pervasive influence of the Western model of education that despite all of Nyerere's efforts to humanize it for the African location, was still based on its top-down philosophy and objectives.

To introduce a first-hand experiential moment, consider this account by one of the authors (C.M.), who is originally from Tanzania. During her first year of graduate school, she took a human resources development course, and one of the assigned readings was an article on Nyerere's "Education for Self-Reliance." The instructor and other students waited anxiously to hear a first-hand viewpoint by someone who had experienced a self-reliant education system. This is what she told them:

> Education for Self-Reliance flopped and everyone in Tanzania hated it. Nyerere had an intimate relationship with Mao Tse Tung. He frequently visited China and was impressed by and decided to embrace Chinese political and economic policies. Nyerere figured out that Tanzania would never progress without Ujamaa and collective farming.

C.M. immediately noticed that her colleagues looked at her as if she was a moron. She asked herself "have I said something wrong about Nyerere?" Clearly, this was troubling, and C.M. decided to read not only the article once again but other writings by Nyerere. This was her first significant attempt to get to know who Nyerere was. Apparently, C.M. wasn't the only one suffering from obliviousness. Indeed, she also recalls what happened when, as a doctoral student at the University of Alberta in Canada, she and others received, with "great sadness, the devastating news of Nyerere's death." Tanzanian students studying at the university arranged a mass service to pay tribute to their former president, who they saw as the heroic father of their nation. One of the students refused to participate. He was bitter about the whole educational project of Nyerere, and resented his policies of Ujamaa, which this student described as not allowing people from his tribe to participate in modern development as they had wished. Ironically, in his isolation and disengagement from the rest of the group, he decided to read about Nyerere, and later confessed, "It has just come to my realization that I never knew what kind of person Nyerere was, and what we have lost as a nation."

The main reason we are relating these personal but important narratives is to highlight how many people, especially the younger generations of Tanzania, either have not been informed or have been misinformed about the noble objectives of Nyerere's educational and social development policies. Of course, it is legitimate to question why many of us had no clue of what kind of person Nyerere was, and how was it that we did not know what he was really doing for Tanzania. Interestingly, Nyerere himself was not oblivious of the forces, whether local or global, he was against. During one of the most active periods of his socialist construction, Nyerere (1976) himself made the following statement: "I am becoming increasingly convinced that we in Tanzania either have not yet found the right educational policy, or have not yet succeeded in implementing it or some combination of these two alternatives" (p. 34).

In terms of the problems with implementation, the zigzagging web of bureaucratic unpredictability that characterizes many projects in both the developing and developed worlds is well known. In the Tanzanian case, specifically with respect to Ujamaa policies, a huge gap existed between the policies and actual implementation. There were at least two categories of implementers: professional civil servants on the one hand, and governing party officials on the other. As generally happens, the relationship between technocrats and party officials was not a healthy one. The majority of the civil servants had benefited from formal schooling up to the university level, and considered themselves as intellectually distinguished from the rest. In addition, this group's thinking was highly influenced by theories of modernization (see Huntington, 1971), and a number of them knew how to directly benefit from their positions of privilege, especially the elite jobs they held in senior government departments. Comparatively, party officials were mostly political appointees and were not as educated (in terms of formal schools and credentials) as the senior bureaucrats.

Considering the demand and expectations created by the policies that Tanzania embarked on, one could easily see how implementers struggled to get things done. For professionals, the struggle was reflected in the conflict between what was stipulated in the new policies and their desire for modern life. The struggle for party officials was of a different nature. They were quite aware that Nyerere was fighting elitism, and to them any professional was considered to be an elite and therefore against Ujamaa. As such, they assumed a self-imposed responsibility of acting as watchdogs for anyone who was against Ujamaa. Aware that they were being watched, the professional cadre, who lacked confidence in their work, were quite insecure, and not even willing to ask questions for clarification. The party officials, who had vested interest in safeguarding their positions, created a false impression by exaggerating records of newly created Ujamaa villages. As a result of these conflicting schemes and interests, different messages were produced for mass consumption. What made things worse was that the strong oligarchic tendencies of many within the party were never successfully contained (Pratt, 1999). For instance, 1968 marked the assertion of the ruling party supremacy, during which lethargic party branches were rejuvenated and new branches were created in workplaces (Samoff, 1990). As a result, the regional and area commissioners, who served as representatives of the president and other party officials, wielded far greater power than one could imagine.

Essentially, policy implementation had three levels: in the lead was the president, constructing ideas for new policies; second were the different implementers, with the duty of delivering messages (translated into programs) to the masses; and the third level was made up of the public, the intended beneficiary of the programs. Considering the low level of understanding among implementers and the nature of the conflict between the professionals and party officials, more often than otherwise, the programs developed for the masses were completely different from what Nyerere had proposed. See Figure 4.2 for a pictorial depiction of the implementation structure.

The primary goal of Nyerere's policies was to bring about social change through the increased awareness of the masses, who were expected to effect such change. However, apparently only Nyerere knew what he was talking about. Given the fact that the socialist transformation the country embarked on was new to the citizenry, one should consider the masses as situational learners. While Nyerere himself was well educated and learned through critical and reflective thinking focusing on important social, cultural, and political issues, the public were never given the opportunity for a similar kind of learning to occur. Moreover, because the implementers (in particular, party officials) controlled the learning environment, they directed learners to do what they thought was necessary to be done. The process that was supposed to be an equalizing process, became, for all intents and purposes, a highly top-down approach of development. With the growth of these problems, people's feelings and doubts about reform added to the implementation slow down. With these weaknesses in implementation leading to negative results, the majority of people in rural areas failed to carry out the orders of the implementers and lost their enthusiasm for Ujamaa.

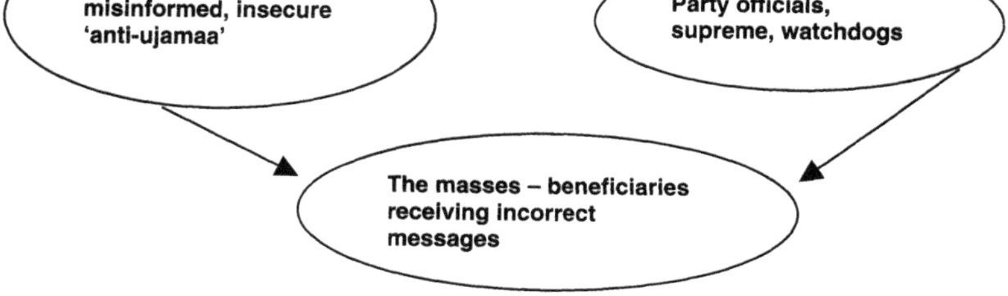

Figure 4.2 Values informally learned combined with academic theories

Did Nyerere Fail, or Who Failed Whom?

Many critics point out that the United Republic of Tanzania's attempts to build a socialist and self-reliant society through political, economic, social, and educational actions largely failed (Kassam, 2000), and that Nyerere's policies were disastrous. In the Arusha Declaration, Nyerere commented on these and similar exhortations. The declaration stated what Tanzania stood for and that the peoples efforts to achieve those goals made Tanzania distinctly Tanzania. He accepted the fact that in trying out anything new, there were bound to be mistakes. In his interview with Bunting (1998) he further clarified, "[W]e cannot deny everything we accomplished. . . . The Arusha declaration and our democratic single-party system, together with our national language, Kiswahili, and a highly politicized and disciplined national army, transformed more than 126 different tribes into a cohesive and stable nation."

In contrast, Pratt (1999) points out Tanzania's political stability, which may not be comparable to any other African state. He argues that for over twenty years, the 1965 constitution, which he considers an extraordinary feat of creative political engineering, provided a largely unchallenged framework within which the citizens of Tanzania arranged their public affairs and enjoyed continuous and stable civilian rule. One would be hard pressed to call that a failure. When Nyerere stepped down, he did not do so because his policies were wrong or because he knew that he was a failure. By all pragmatic accounts, and to disavow the travails of any generosity that might be dispensed on the platform of the dominant forces' snobbish gentility, Nyerere was anything but a failure. And, if there is any space for describing failure, it should only have select temporal connotations in that perhaps the program was ahead of its time, and that there should have been enough preparation to neutralize the selfish aims of some implementers (McHenry, 1994) and to thwart the global project of neoliberalism, which did everything it could to destroy Nyerere's policies and programs. To the counterpoint of why Nyerere did not do the required homework to overcome these issues, the simple answer is that he was diligently working with the contextual endowments he could procure for himself and his compatriots. He was not willing, we should know by now, to surrender to the predatory world forces that were determined to punish any political or social entity that did not toe their line.

It is also important to note that Nyerere's policies were geared to control elitism, a class that emerged immediately after independence and, undoubtedly, during his reign, the gap between the haves and have-nots was one of the smallest in the world. In today's Tanzania, with unfettered laissez-faire economics, the only social development one can speak of is the tremendous wealth that has been created for a tiny elite. The few stupendously wealthy in the country, along with global capitalists, would, of course, admire this triumph of the monetarist project complemented by the deliberate shelving of Nyerere's policies.

As Pratt (1999) notes, Nyerere's central aim of not surrendering control of Tanzania's socioeconomic development to international capitalist interests or international agencies dominated by the major industrialized states reflected not only nationalist aspirations, but also a profound belief that integration into the international economic system would bring little advantage to the world's poorest countries, including his own. This would be so if the poor countries were, and they seem to be, unable to manage skillfully and selectively their relationships with the major capitalist countries. This remained a central concern of Nyerere throughout his life. Today, foreigners, for example, are investing in the country's natural resources such as mining, extraction of gas, and other service-based resources like tourism. The Tanzanian government has signed a good number of agreements with these foreign investors; however, the citizens of Tanzania remain largely uninformed about the investment agreements and all the benefits accrued from these investments. How are these economic activities responsive to peoples' needs? Do the citizenry have the control of these economic activities? Do they know what such investment entails and what benefits will come to the ordinary man and women in the street and in the village? Is there a provision in the Tanzanian investment policy for people to be informed about and possibly participate in these agreements? The answer to all of these questions is a resounding no.

The consequences of these policies are obvious and the majority of people in Tanzania, particularly the rural and urban poor, are learning to accept this new reality. On the one hand, people living in those areas where there is investment are helplessly watching airplanes ship away tons of fish from Lake Victoria and foreign companies ship away sand from the mines. On the other hand, people are dying during childbirth because they cannot afford to pay for hospital services. During Nyerere's reign, there were health centers in almost every village and almost every citizen was entitled to basic education and health services. Today, many in the country cannot afford one meal a day, cannot afford to pay for treatments in hospitals, cannot afford higher education for their children, and cannot find a job after training. Indeed, the saying that "One does not know freedom until she/he has lost it" would be relevant here. The majority of poor people in Tanzania have undergone a long learning and reflective journey with the rich experiences of colonization, independence, the Arusha Declaration, the policies of Ujamaa and, finally, the experiences of the free market economy. Let the question be directed to these people as to whether Nyerere's policies were disastrous for Tanzania's rural and the urban poor, who make up over 80 percent of the population? The answer should be obvious: Mwalimu, as an educator, politician, and policy maker practically understood the bitter inheritance, in terms of education and human progress, of colonialism, and decided to challenge headlong the dangers and greed of global capitalism. It is our understanding, therefore, that Nyrerere's policies and programs would have achieved so much better for the people of Tanzania, and undoubtedly for the people of Africa. Instead, the current projects of globalization and American democracy have created a situation where the pauperization of African vis-à-vis the rest of humanity is now a *fait accompli.*

Conclusion

By the late 1960s, when Nyerere took over the leadership role from British rule, Tanzania was one of the world's poorest countries. Nyerere, steered a difficult course during his presidency and committed himself to broad-based social and economic development anchored in significant social structural change. Instead of picking up from European colonialism and simply following the former master's colonial education and related resources management arrangements, Nyerere, quite boldly and with thick philosophical and cultural convictions, decided to attempt a different course for postcolonial Tanzania. Nyerere, as a teacher as well as a politician, was fully committed to education. He theorized about it and put into practice new programs of education for self-reliance that were to serve, as he put it, not the aims of the country's colonial past, but the immediate needs of the people. This was not, of course, easy. Transforming a whole educational and social development paradigm with limited resources, with less commitment from those who are expected to implement the programs, and with the inimical incessant onslaught of global capitalism was undoubtedly too much to face, and at least the combination of some of these seem to have derailed the practical side of Nyerere's Ujamaa projects. Nyerere, of course, understood these challenges, and up to his death believed that he was on the right side of the real needs and aspirations of the Tanzanian and African peoples. Nyerere's programs of adult education were especially important in that they were intended to uplift the lot of the largely illiterate adult population, without which, Nyerere rightly believed, the country would not, in advancement terms, move forward. Still, the Tanzanian experiment offered a number of important insights into possible educational, economic, and social alternatives to capitalist development, alternatives that are still relevant for Third World countries. Nyerere willingly retired in 1985, leaving the country to enter its free market era under the leadership of Ali Hassan Mwinyi. There are speculations that he was unwilling to lead Tanzania using an economic model he did not believe in. Whether it is a pragmatic assessment or not, it might have been useful for the people of Tanzania to experience a change, a change that would by now have exposed the realities of the capitalist projects that are creating, in Tanzania and elsewhere in the continent, a tiny group of haves who are "developing" their countries at the expense of the overwhelming majority.

References

Abdi, A. (2002). *Culture, education and development in South Africa: Historical and contemporary perspectives.* Westport, CT: Bergin & Garvey.

Abdi, A. (in press). Democratic development and prospects for citizenship education: Theoretical perspectives on sub-Saharan Africa. *Interchange: A Quarterly Review of Education.*

ANC. (1999, October 14). A people's hero: African National Congress Statement on the death of "Mwalimu" Julius Kambarage Nyerere.

Bunting, I. (1998). Interview with Julius Nyerere. Dar-Es-salaam, 1973

Coffield, F. (Ed.). (2000). *The necessity of informal learning.* Bristol: The Policy Press.

Freire, P. (1985) *Education for Critical Consciousness.* South Hadley, MA: Bergin & Garvey

Huntington, S. (1971). The change to change: Modernization, development and politics. *Comparative Politics, 3*(3), 283–322.

Kassam, Y. (2000). *Julius Kambarage Nyerere.* Paris: UNESCO International Bureau of Education.

Mayo, P. (2001). Julius K. Nyerere (1922–1999) and Education—A tribute. *International Journal of Educational Development, 21*(3), 193–202.

McHenry, D. (1994). *Limited choices: The political struggle for socialism in Tanzania.* Boulder, CO: Lynne Rienner.

Merriam, S.B., & Cafarella, R.S. (1999). *Learning in adulthood: A comprehensive guide.* San Francisco: Jossey-Bass.

Mezirow, J. (1990). *Fostering critical reflection in adulthood.* San Francisco: Jossey-Bass.

Nyerere, J.K. (1968). *Freedom and socialism.* London: Oxford University Press.

Nyerere, J.K. (1976). "Development is for man, by man, and of man": The declaration of Dar-es-Salaam. In B. Hall & J. Roby Kidd (Eds.), *Adult learning: A design for action* (pp. 27–36). Toronto: Pergamon Press.

Nyerere J.K. (1979). Education never ends. In H. Hinzen & V.H. Hundsdorfer (Eds.), *The Tanzanian experience: Education for liberation and development.* Hamburg: Evans Brothers.

Polanyi, M. (1962). *Personal knowledge: Toward a post-critical philosophy.* Chicago: The University of Chicago Press

Polanyi, M. (1966). *The tacit dimension.* London: Routledge & Kenya

Polanyi, M. (1968). *Personal knowledge: Towards a post-critical philosophy.* Chicago: The University of Chicago Press.

Pratt, C. (1999). Julius Nyerere: Reflections on the legacy of his socialism. *Canadian Journal of African Studies, 33*(1), 137–152.

Rahnema, M. (1976). Education and equality: A vision unfulfilled. In B. Hall & J. Roby Kidd (Eds.), *Adult learning: A design for action.* Toronto: Pergamon Press.

Samoff, J. (1990). Modernizing a socialist vision: Education in Tanzania. In M. Carnoy & J. Samoff (Eds.), *Education and social transition in the Third World.* Princeton, NJ: Princeton University Press.

Smith, M. (1998). (Ifed): Julius Nyerere, lifelong learning and informal learning education. Accessed www.infedfromJune15,2006,org/lifelonglearning/b_edcom.htm

CHAPTER 5

Globalization, Dispossession, and Subaltern Social Movement (SSM) Learning in the South

Dip Kapoor

No agrarian reform is acceptable that is based only on land redistribution. We believe that the new agrarian reform must include a cosmic vision of the territories of communities of peasants, the landless, indigenous peoples, rural workers, fisherfolk, nomadic pastoralists, tribal, afro-descendents, ethnic minorities, and displaced peoples, who base their work on the production of food and who maintain a relationship of respect and harmony with Mother Earth and the oceans. (Statement of the International Planning Committee on Food Sovereignty, cited in Via Campesina, 2006)

This statement by a global coalition of over 500 rural organizations, including the influential Via Campesina (one of the world's largest peasant, indigenous, and landless people's network), suggests that subaltern[1] social groups in Africa, Asia, and Latin America (pejoratively referred to as the Third World and alternatively as the global South) continue to assert their claims to both modes of production and modes of meaning making, in conjunction with the increasing penetration of a globalizing (imperialist) capitalist colonization of material and cultural space (referred to in this chapter as neoliberal globalization) by the countries of the TRIAD, including North America, Europe, and Japan. While Frantz Fanon (1963) observed that Europe was "literally the creation of the Third World," an opulence that was fuelled by "the sweat and the dead bodies of Negroes, Arabs, Indians and the yellow races" (p. 76), the postindependence dependent-development project and the current wave of 1980s neoliberal globalization, while continuing to extend the reach of Euro-American imperialism, have simultaneously created a transnational elite and consumer class (in parts of Asia, for instance) that cuts across space-race boundaries, a development that has been enabled by the continued displacement and dispossession of subalterns and ecological ethnicities in the South.

Neoliberal globalization and the process of modern development and dispossession continue to be disrupted by numerous local and increasingly global subaltern assertions to "deflect globalization's reinvention of colonial processes" while being "within, besides and against colonization" (Barker, 2005, p. 20). Despite the formation of new social institutions in rural and urban life in the global South through centuries of direct colonial occupation and postindependence neocolonialism (through dependent-development or active development of underdevelopment in the South), and the current neoliberal globalization process whereby the power of the United States and leading European states is activated by their control of the international financial institutions or IFIs, such as the World Bank, International Monetary Fund [IMF], and World Trade Organization [WTO]) to support the interests of 37,000 transnational corporations (TNCs) (85 percent of whom are headquartered in these countries) in opening up Southern markets through privatization and trade and investment liberalization (Petras & Veltmeyer, 2001), subaltern collectivities in the South have "continued to exist vigorously and even develop new forms and content" (Ludden, 2005, p. 100). These formations include today's contemporized and increasingly resilient Southern subaltern social movements (SSMs), sociopolitical formations marked by their perennial political presence and obstinacy (refusal to disappear?) in the face of these repeated colonizations (Martinez-Alier, 2005; Moyo & Yeras, 2005; Polet & CETRI, 2004; Sousa Santos, 2007).

In this chapter, I attempt to establish the continued political and academic significance of SSMs in the South as contemporary sites of social praxis, adult learning, and popular education by giving definition to these formations and enhancing their visibility (located in the margins in the South, they often become or are made invisible) within critical/popular adult education/learning (Allman, 1999, 2001; Cunningham, 1998, 2000; Foley, 1999; Holst, 2002; Kane, 2000, 2001; La Belle, 1986; Mayo, 2004; Torres, 1990) and scholarship preoccupied with social movement phenomena and the prospects for social change in the global South (Alvarez, Escobar, & Dagnino, 1998; Escobar & Alvarez, 1992; Ray & Katzenstein, 2005; Starn, 1992; Wignaraja, 1993). This will be accomplished by (a) a brief analysis of globalization and dependent–development/modernization as processes of politico economic and cultural imperialism (global and national capitalist colonizations); (b) sketching the contours of SSMs and pointing to these formations across the South in relation to processes of dislocation, dispossession, and the repeated colonization of subalterns via development and globalization projects; and (c) discussing the political and academic significance of these movement sites for critical/popular adult education and movement learning in the South, with reference to the author's current research in "Learning in *Adivasi* (Orginal Dweller) Movements in India."[2]

Globalization and Dispossession of Subalterns in the South

While the critique of globalization and its implications for subalterns is often preoccupied (as is the analysis presented in this chapter) with its economic vector or the global spread of capitalism, globalization is a multidimensional process (including economic, political, cultural, and technological aspects defined, legitimated, or informed by the culturo-political principal/seed of universalism. Political philosopher John Gray (1997) defines *globalization* as "only a perverse and atavistic form of modernity—that is, roughly, of nineteenth-century English and twentieth-century American economic individualism—projected worldwide" (p. 183); a process that is also driven by the "metaphysical faith that local western values are authoritative for all cultures and peoples," a notion in turn derived from "the Christian commitment to a redemption for all humankind and in the Enlightenment project of progress towards a universal human civilization" (p. 158). Subaltern groups are subjected to globalization as Western universalism, embracing both economic and cultural dimensions of imperialism (material and cultural displacements and dispossessions), creates a "vicious universalism" (Tomlinson, 2001, p. 48) with varied critical implications for various universalizing projects including neoliberalism, socialism, liberal human rights politics, some forms of environmentalism, and feminism.

With this notion of so-called vicious universalism as the link between economic and cultural globalization in mind, globalization understood as imperialism (an advanced strain of capitalist colonialism) is by such definition a "prescriptive project" which continues to serve the interests of national and transnational capitalist classes and is subsequently "contestable as a social construction," as opposed to being an "incontestable natural process" or "objective description of an inevitability" as advanced under the hegemonic propensities of neoliberal ideology

(Bowles, 2007, p. 181; Petras & Veltmeyer, 2001). This process includes ensuring the uninterrupted exploitation of subaltern people's labor, land, forests, and water by using free trade to open up Southern economies and spaces and subsequently continuing to perpetuate global, regional, and national inequality and hierarchies of political, economic, and cultural power. While the imperial expansion of Europe into Asia, Africa, and the Americas restructured the economies of these colonized regions and produced the requisite imbalance necessary for the growth of modern European capitalist colonialism (Chinweizu, 1975; Galeano, 1973; Rodney, 1972), post-war twentieth-century development pertaining to these newly independent countries ensured a process of continued colonial advantage via dependent–development and the active development in underdevelopment of colonized regions (Mason, 1997). This, of course, happened with the help of an emerging and modernized Third World elite now complicit in a process of Euro-American and internal colonization of subalterns and other subordinate social classes—what Gandhi referred to as "English rule without the English" or "the tiger's nature without the tiger" (Parel, 1997, p. iv) in the Indian context and what Fanon and Cabral identified for Africa in the 1960s. In the later part of the twentieth century, globalization has fostered the development of a transnational capitalist/consumer class (Sklair, 2001) and an increasingly globalized division of labor with corresponding circles of exclusion of vast numbers from the market (Hoogvelt, 2001). This process is beginning to undermine the previously racialized division of marginalization and colonial privilege monopolized by the Europeans as Asian capitalists and consumers now begin to partake in the spoils at the colonial table, mostly at the expense of subaltern groups both within and beyond their national containers.

Large dams are but one example of this parasitic process of dispossession of subalterns while mining, oil and gas, forestry, agribusiness, tourism, highways, military installations, and game sanctuary/reserves, to name a few possibilities, only compound this picture. In the Final Report of the World Commission on Dams (November 17, 2000), a commission composed of representatives from all sectors, it was estimated that some 80 million people have been displaced by dams alone between 1950–1990 (p. 104) and that "the direct adverse impacts of dams have fallen disproportionately on rural dwellers, subsistence farmers, indigenous peoples, ethnic minorities and women" (p. 124). Dams in China and India (financed by IFIs and power sector TNCs with financial interest in such mega projects) alone account for between 26 and 58 million people displaced; the Three Gorges Dam (China) and the Narmada

Valley Project (India) together displaced an additional 4 to 7 million people at the turn of this century. Large dams have adversely affected flood plain agriculture, fisheries, pasture, and forests "that constitute the organizing element[s] of community livelihood and culture" in Africa (p. 83), while dam-induced flooding has affected the lives of some 65 million people between 1972 and 1996—"more than any other type of disaster including war, drought and famine" (p. 14).

Despite the emerging trend of pan-global dispossession/colonialism and the linkage between inequality–prosperity (Bond, 2006; Chossudovsky, 2003; Petras & Veltmeyer, 2001), the geographical (and concomitantly racial) division of North/South is simultaneously persistent (putting aside, for the moment, the postcolonial and poststructural/developmental angst of culturalists understandably concerned about the implications of such terminology/signification vis-à-vis perpetuation of ideological, mental, and subsequently material colonizations) and cannot be dismissed with rhetorical flourishes if we are to account for the colonial import of the historical and contemporary processes of neoliberal globalization which are exacerbating North-South regional/national colonial impositions (Arrighi, 1991) while creating new prospects for transnational processes of colonization that cut across geographic and racial borders.

What this means for subaltern groups is that neoliberal globalization has effectively multiplied the tentacular possibilities of colonial control of these social groups in a number of ways. First, it has widened North-South regional divides by augmenting the neocolonial controls initially implemented through the post-independence dependent–development project.[3] Second, by creating the conditions for continued and increasing internal inequalities within nation states, it has heightened prospects for "internal colonization" (Mignolo, 2000). For example, the efforts of local/national industrial capitalists and the growing national consumer class, with their sites trained on subaltern water, land, labor, and livelihoods—allegedly for national and global progress and modernization, have encouraged development-led displacements and dispossession of subaltern rural/forest communities (multiple displacements are not uncommon). And, third, globalization has expanded transnational processes of penetration; for example, TNCs can now own and operate mines and extractive industries within national regions and subaltern spatial locations entirely for export production, paying little heed to local environmental impacts or the sociocultural and material fall-out from the "development-triage" of subaltern lives.[4]

In the case of India's 80 millon plus Adivasis (referred to as tribals, Scheduled Tribes, or STs under the Constitution), for instance, Behura and Panigrahi (2006) conclude that the postindependence scenario has witnessed the continued victimization of Adivasis through a "systematic process of exploitation, which has marginalized and impoverished them. . . . State policies on land and land-based resources instead of encouraging the tribals have depressed them and opened up the tribal economy for others to exploit" (p. 209). Unsurprisingly, Adivasis constitute just 8 percent of the population yet account for 40 percent of displaced persons (Fernandes, 2006); furthermore, 95 percent of mining activities today are on Adivasi land alone, since the implementation of the New Economic Policy (neoliberal marketization and privatization of the Indian economy) of 1991.

Promoting market access has become the key neoliberal response to eliminating poverty. This, in turn, entails disciplining the poor, who are "presented as inhabiting a series of local spaces across the globe that, marked by the label 'social exclusion,' lie outside the normal civil society. . . . Their route back . . . is through the willing and active transformation of themselves to conform to the discipline of the market" (Cameron & Palan, 2004, p. 148). Sadly, the World Bank's latest attempt to address poverty, the Poverty Reduction Strategy Papers (PRSPs) of 1999, a refashioning of the discredited SAPs, elicits the following observation from one analyst: "Most PRSPs, for all their emphasis on 'pro-poor' growth, do not include decisive measures to redistribute wealth and promote equality. Land reform, for example, is studiously avoided in the majority of plans, despite its importance for the reduction of rural inequality and poverty" (Abrahamsen, 2004, p. 185). A statement from the Thai Assembly of the Poor (a chapter in Via Campesinaa, 2005) notes that, as a result of such policies, "major production resources such as land were lost by the small farmers (over 1.5 million farming households either became landless or did not have enough farmland between 1995 and 2003). Their rights to use of resources related to production, such as water, forest, local genetic, and coastal resources were also infringed on" (pp. 25–31).

Whether in the days of British colonialism in India, or the multiple trajectories of European colonialism in the South, or in relation to colonialism through regimes of capitalist exploitation established via dependent–development and neoliberal globalization, there are numerous examples (historical and contemporary) of SSM assertions and/or spontaneous struggles (Martinez-Alier, 2005; Moyo & Yeras, 2005; Peet &

Watts, 2004; Polet & CETRI, 2004; Sousa Santos, 2007) in relation to material and cultural displacement and dispossession and attacks on subaltern well-being. In terms of a working definition of *social movement* being employed in relation to SSMs here, some defining and connected elements could include (Kapoor, 2007, p. 19): (a) movement as indicative of an articulation of concerns/issues (for example, Adivasi struggles around "own ways," land, water, and forest and cultural chauvinisms, racism, and discrimination or what Mignolo [2000, p. 7] refers to as "the coloniality of power" and the "colonial difference"); (b) movement as defined by the maturity and growing unity of an organized presence/vehicle for such articulations (for example, emergence of movement organizations) that engages a critical mass of people with a like-concern for core and evolving sets of "movement issues"; and (c) movement as organized action directed at oppositional (colonizing) social structural and institutional forces that "give cause" for such movements in the first place. Alternatively, and additionally, synthesizing European and North American literature on the subject, Della Porta and Diani (1999) identify four movement characteristics as "informal interaction networks; . . . shared beliefs and solidarity; . . . collective action focusing on conflict; . . . use of protest" (pp. 14–15).

Subaltern Social Movements in the South

From flagship subaltern movements in the South, including the Ejército Zapatista de Liberación National (EZLN) or Zapatistas (Mexico), the Narmada Bachao Andolan (NBA) and Chipko movements (India), the various indigenous movements of the Mindanao Islanders and the anti-dam movement of the Igorot (Philippines), the Dayak Indian movement (Indonesia), the landless people's movement or O Movimento das Trabalhadores Rurais Sem Terra (MST) (Brazil), the Coordinadora (Bolivia), the Ogoni movement (Nigeria), the Greenbelt women's movement (Kenya), or the various struggles and movements of the San peoples of the Kalahari (Southern Africa), to numerous lesser known struggles of subalterns[5] directly affected (ecologically, economically, politically, socioculturally, and spiritually) by the processes of transnational capitalist colonial exploitation, SSMs have and will continue to play a significant role in the political history of resistance and resilience (Martinez-Alier, 2005; McNally, 2002; Moyo & Yeras, 2005; Polet & CETRI, 2004; Sousa Santos, 2007) in the face of tremendous odds.

The Politics of Movement Representations and Realities: Locating and Disembedding SSMs from NSM-OSM Taxonomizations

Social movement (Carroll, 1997; Cohen & Rai, 2000) and related critical adult/popular education/learning scholarship (Holst, 2002) and the question of movement teleologies are mostly preoccupied with the Old Social Movement (OSM; labor vanguard politics of the material in terms of establishing control over the means of production and the distribution of economic wealth in industrial and postindustrial society) and New Social Movement (NSM; a civil societarian, predominantly middle-class politics of culture and identity), or the Global Social Movement (GSM; Cohen & Rai, 2000) distinction, none of which prove to be appropriate analytical or praxiological categories when it comes to SSMs in the South. The OSM/NSM categorization has generally been adopted when considering the development of movement taxonomies in the South (Alvarez et al., 1998; Escobar & Alvarez, 1992; Wignaraja, 1993), despite their combined disregard for the issues, concerns, and historical trajectories and experiences of subaltern groups and their movement specificities in the South. This is hardly surprising as Euro-American scholarship on movements is tied to postindustrial society and is largely consistent with the preoccupations of modern-developmentalist and globalist ideologies. In fact, peasant, indigenous, rural existences and the well-being of ecological ethnicities in the margins of the South have been fed to the combined harvesters and threshed in to the dustbins of progress, both in relation to theoretical analytics and categories and as social and living struggles on the ground.

In the interests of a politics of and at the margins, SSMs need to be salvaged from their relative invisibility and from recent attempts to include them in NSM/GSM categories. SSM teleologies and definitions need to be understood on their own terms, as opposed to being subjected to a logocentric analysis relative to antiglobalization and the teleologies of various NSM/GSMs or as compartmentalized submovements within what some refer to as the global human rights movement or the global ecology movement, as Cohen and Rai (2000) attempt to do in the case of indigenous SSMs. Such clarity would arguably also benefit from a disembedding of SSMs from the "globalization from below" category commonly framed in the discussion on the politics of globalization. In the interest of resuscitating SSMs in relation to a politics of representation and probable onto-epistemic colonizations of SSM locations and their distinct and urgent politics, it is necessary to identify some distinctions (at the risk of over simplification) as it very well might not be an exaggeration to suggest that there are as many SSM movement teleologies as there are subaltern groups or SSMs or subaltern contextual situations. Possible points of distinction between SSMs and the NSM/OSM categorization (I will refrain from elaborating on the distinctions/ intersections among NSM/OSM categorizations that have been more than adequately dealt with in Carroll, 1997, or Holst, 2002, for example) arguably include but are not limited to some of the following six dimensions (the inapplicability of these dimensions would depend on the specific subaltern group and the specificities of their historical, politico-economic, and cultural location).

1. Agents of SSMs, as defined by the Via Campesina (2006) and a coalition of 500 rural organizations across the globe, include peasants, the landless, indigenous peoples, rural workers, fisherfolk, nomadic pastoralists, tribes, afro-descendents, ethnic minorities, and displaced peoples united by physical and spatial locations in rural forest and water geographies with attendant implications for modes of production such as small-scale food producers and living economies with a relatively smaller ecological footprint.
2. SSMs are being subjected to the loss of the means to reproduce their material existence as rural ecological resource economies via a process of colonial dispossession from land, water, and forests perpetuated by feudal elites and national and transnational neoliberal state-market forces (i.e., they face a continued collective experience of the "coloniality of power" as physical and material space is usurped). SSMs, therefore, must struggle against material colonizations and dispossession by globalist and developmentalist projects.
3. SSMs have a unique mythico-religious basis. Their social groups and ecological ethnicities have close spiritual and religious ties to their historic sense of physical and existential location. Place takes on both material (as discussed above) and spiritual-religious connotations (material and nonmaterial ontological understandings are embraced). SSMs can be also defined, distinct from NSM-critiques, as being essentialist, rooted, traditional, exclusionary/ethnocentric, and patriarchal (from an SSM view, these critiques are also examples of the continued collective experience of the coloniality of power as spiritual-cultural and religious space is racialized, stigmatized, and seen to be in need of modern treatment for "backward" ways of being). SSMs struggle against cultural, religious, and spiritual colonizations and for the autonomy to determine how to address internal schisms, oppression, and inequalities (critical traditionalism and sociocultural renewals).

4. SSMs and their agents are motivated by the direct and immediate material impacts of colonial developmentalist or globalist displacements and dispossession and the resulting suffering (e.g.) dam displacements, ecological destruction, loss of existential spaces, and resulting poverty and hunger). "Environmentalisms of the poor" (Martinez-Alier, 2005) recognize this dimension and the precariousness of subaltern existence in neoliberal times, as compared to middle and upper-class environmentalisms of the telescopic/empathetic variety, whereby NSM agents are not necessarily directly impacted by, for instance, ecological destruction of a rainforest, while indigenous subalterns living in those forests face forced displacement. Often ecological NSMs and GSMs contradict SSM politics, as the former speak from the relative security of their remote urban locations (consuming resources "here" while aiming to "protect nature over there," while disregarding the contradictory plight of subalterns "in nature over there"). Meanwhile, indigenous leaders and communities are the miner's canary for humanity and face the invidious task of choosing development and thereby being condemned by green social movements and some of their own community for "saving the village by destroying it" (their claim to indigeniety is also subsequently delegitimated), or refusing cooptation or joint ventures and thereby adding to their lack of autonomy (Cohen & Rai, 2000, p. 25).

5. SSMs Sociopolitical location outside of modernist society-polity configuration results in dubious status, at best, in terms of national citizenship. SSMs are often compelled to exert political pressure from the margins in relation to the state (e.g., political party process and cooptation and vote-banking tendencies toward subalterns), the market (e.g., transnational corporate colonizations of material space), and civil society (e.g., against depoliticizing service-oriented NGOs acting as global soup kitchens attempting to manage subaltern discontent germinating from neoliberal state-market processes of subaltern displacement or GSM coalitions that fail to give due recognition/space to SSM teleological priorities, as has been the case with numerous indigenous SSMs, North or South).[6]

6. SSM movement teleologies and political strategy, given their location outside the state-market-civil society nexus, separate SSM articulations from NSM/OSMs in some of the following ways. (a) SSMs use a complex blend of struggles around (2) and (3) (against material and ideo-cultural capitalist colonizations and globalist imperialism in the interests of any number of possibilities, such as complete secession/ independence (separatist SSMs in a minority of cases), sovereignty,[7] plurinationalism, self-determination, autonomy, inclusion, or opportunity structures for integration. (b) The strategic necessities that dictate coali-

tional politics may lead various SSMs alternatively to political parties, NSMs/OSMs and GSMs or partnering among SSMs themselves in order to achieve the preceding aims of a given movement. (c) Taking a cue from Via Campesina (2006), for SSMs the demand for sovereignty means "recognizing the laws, traditions, customs, tenure systems, and institutions, as well as the recognition of territorial borders and the cultures of peoples," that is, the appeal to territory represents "a substantive demand to affirm citizenship as a basic national human right, but also as a vehicle for the sovereign right of minorities, creating pluri-national states as a precondition for protecting and sustaining peasant spaces to overcome the crisis that is neoliberal development" (McMichael, 2006, p. 479). Perhaps, as Sheth (2007,) observes, some SSMs might well be engaged in a process whereby "they now seek to change the power relations on which the conventional model of development is premised" (p. 15) or are attempting to transform the state/development project by, for example, Via Campesina's endorsement of the right of states to "define, without external conditions, their own agrarian, agricultural, fishing and food policies in such a way as to guarantee the right to food and other economic, social and cultural rights of the entire population" or by their insistence on the "rights of peoples, communities and countries to define their own agriculture, labor, fishing, food and land policies which are ecologically, socially, economically and culturally appropriate to their unique circumstances" (quoted in McMichael, 2006, p. 479). (d) SSMs increasingly relate the globally debated issues such as feminism, ecology, and human rights (essentially subjects that have not emerged from subaltern communities) to the economic, social, and cultural specificities of their own locations, engaging in strategic deployments of these rights and vocabularies (taking advantage of political spaces opened up by NSM/OSM and civil society dissidence in imperial societies) in developing counterhegemonic and antihegemonic possibilities (Kapoor, 2008).

This partial and initial attempt to define SSMs as movement formations that are distinct from NSM/OSM/GSM categorizations is neither exhaustive nor is it intended as a foundational statement on these complex movement identities and dynamics. such categorization is prompted by the need to acknowledge and recognize subaltern politics and political aspiration as unique formations addressing continued colonizations, including those by allegedly progressive modern political forces and movements (civil society). If the above distinctions are of any analytical and praxiological import, attempts to appropriate SSMs into preexisting Euro-American perspectives on movements and radical adult education

(Marxist appropriations of SSMs via peasantization/homogenization of all subalterns into the historical revolutionary project of OSMs or civil societarian NSM/GSM appropriations of the same into a middle-class urban consumerist politics of identity, placelessness, individual rights, and telescoping ecological activisms from within the confines and comforts of urban market locations are both embedded exclusively in Enlightenment onto-epistemic and axiological origins)[8] will amount to a continued colonization of SSM interpretations and praxis at worse or, at the very least, end up caricaturizing the teleological complexities at play in subaltern assertions. These socio-theoretic interpretations and impositions and the aforementioned dimensions of SSMs have important implications for how researchers and movement participants go about engaging with movement praxis for social change and for purposes of SSM-informed academic understanding and analysis pertaining to popular education regarding social movements, as in the extensively grounded work of Liam Kane (2000, 2001) with the landless workers movement in Brazil and related theoretical and conceptual formations introduced by Griff Foley (1999), or pertinent frames of reference from the sociology of adult education and social activism (Cunningham, 2000).

SSM Learning in the South: Emerging Analytical and Praxiological Considerations from Adivasi Movements in India

Keeping in mind the impositions of neoliberal globalization and development as partially linked processes of continued colonization of subaltern social groups and the related positioning of distinct (from NSMs and OSMs) SSM formations, the following observations concerning possible analytical categories and praxis-oriented possibilities have emerged and continue to emerge from researching movement learning with predominantly Kondh Adivasis and Dalits (state-classified scheduled castes or "untouchable caste groups") and their SSM organization, the Adivasi-Dalit Ekta Abhijan (ADEA) in eastern India (with a Kondh-Saora-Dalit movement mobilization of approximately over 21,000 people located in 120 villages) since 2006. My relationship with the communities of the ADEA goes back to the early 1990s and includes several participatory action research and/or ADEA organized initiatives pertaining to food, land, forest, and water-related concerns that partially define the ADEA movement aspirations. The purpose of this exposition is to identify useful dimensions of movement learning research (details concerning the participatory methodological orientation to this research can be ascertained from Kapoor, 2007).

The purpose of my research was to determine the aims of the ADEA movement and related issues being faced by Adivasis in this context; how learning contributes toward defining ADEA movement goals and their subsequent achievement; and how various knowledge engagements contribute to learning in movements. In keeping with Castells' (1997) suggestion that "social movements must be understood in their own terms: namely, they are what they say they are," and that "their practices (including discursive practices) are their self-definition" (pp. 69–70), self-definition of the movement by the ADEA has been taken seriously. While this is a context-specific discussion, there is little reason to abstain from making pertinent extensions to movement learning in other SSMs in varied contexts in the countries of the global South while respecting the self-definition and integrity of each movement.

Movement learning (informal and incidental) and popular education dimensions (as organized, purposeful, and often engaged by movement leadership and others) to consider for research/praxis) include macropolitico-economic and socio-cultural issues, micropolitical and intra-communal issues, learning and knowledge processes, and catalytic validity and opportunities for research-instigated movements. These are interactive-interlocking dimensions as opposed to being discrete and ordered, that is, macro-micro political linkages are more the case than not, even though the two dimensions are presented here in apparent isolation.

Macro Politico-Economic and Sociocultural Dimensions

> We are the *mulo nivasi* (root people) and the people who dominate us, as history has taught us, came here 5,000 years ago. . . . Today the *sarkar* (government) is doing a great injustice *(anyayo durniti)* . . . and the way they have framed laws around land-holding and distribution, we the poor are being squashed and stampeded into each other's space and are getting suffocated *(dalachatta hoi santholito ho chonti)*. This creation of inequality *(tara tomyo)* is so widespread and so true . . . we see it in our lives and this is the root of all problems and we hold the government responsible for this . . .
>
> They tell us they want to make machines and industries for themselves. To do this they are doing forceable encroachment of our land—they are all over our hills and stones . . . we have become silent spectators *(niravre dekhuchu)* to a repeated snatching away of our resources . . . whenever we have asserted our land rights, we have been warned by the upper castes, their politician friends and the wealthy and have faced innumerable threats and retaliations. The *ucho-barga* (dominant castes and classes) will

> work to divide and have us fight each other till we are reduced to dust *(talitalanth)*. (Kondh Adivasi man, ADEA village)

ADEA movement learning, as is made evident by this narrative at a Kondh village gathering to discuss the ADEA analysis of Adivasi situation issues, provides some indication that macropolitico-economic understandings (and related issues) shape movement learning, including movement aims to address these concerns. Additionally, the naming of key agents and social vectors such as the "state," the "wealthy," "upper castes and classes," and "the moderns/industrializers" provide important pointers to further avenues for exploration to help shed light on movement issues, purposes, and related learnings. Such learning is also historical and alludes to old and new colonialisms alike, while foregrounding the "coloniality of power" of caste, class, racism, urban-rural divides, and ascribed status as being inferior along with clear descriptions of the levers of power point to a macropolitico-economic and sociocultural map of the process and agents of colonial domination:

> We fought the British thinking that we will be equal in the independent India . . . but the *savarnas* (upper castes) and the rich people have controlled *(akthiar)* the land, including *Adivasi* land. (Kondh Adivasi elder)
>
> They have the power of *dhana* (wealth) and *astro-shastro* (armaments), they have the power of *kruthrima ain* (artificial laws and rules)—they created these laws just to maintain their interests . . . and where we live, they call this area *adhusith* (Adivasi or pest-infested), we are condemned to the life of *ananta paapi* (eternal sinners), as *colonkitha* (dirty/black/stained), as *ghruniya* (despised and hated). (ADEA male leader)

Agents and sociopolitical vectors of political, economic, and cultural power and the relations of power with Adivasis and Dalits that have emerged as key macro dimensions of movement learning and that have guided explorations into what shapes movement learning (the macrological territory) include the following. The neoliberal state (post-1991 liberalization) and specific elements of the bureaucracy that shape Adivasi lives on a regular basis (e.g., agents of the state such as the forest department, the revenue department, public works, judiciary, and law enforcement). Each relationship is excavated and informs movement learning content and praxis for redress and change. The feudal/caste elite (e.g., absentee landlords, landed farmers, and money lenders) and industrial (particularly, in this region, mining) TNCs and their role in material dispossession and caste relegations that perpetuate a culture of subservience and self-denigration. Agents of civil society, including funding agencies or international NGOs and state level NGOs and religious charities, that promote service-oriented dependent relationships which have made it increasingly difficult for SSMs to politicize their constituencies utilizing colonial politico-economic and cultural critique/learning (as NGOs dole out much needed goods and services that silence a culture of assertion aimed at state provision, as per their rights as citizens, and/or return of Adivasi space). And the party-political groups and formations of various shapes and ideologies, which include a virulent strain of Hindu fundamentalist politics in this region.

Micro-Political and Intracommunal Dimensions

Similarly, intracommunal politics have emerged as a key dimension of movement learning and for guiding research into the micropolitics that shape this process. Unlike macrodimensions, micropolitical dimensions are predominantly internal to the communities and their immediate physical location. Furthermore, micropolitical dimensions are treated, for the most part, by the ADEA as secondary to the collective struggle being waged in relation to the politics "external to and directed at subalterns," with the exception of the micropolitics of Adivasi (tribe)-Dalit (caste) unity, which is also implicated in the macro-politics aimed at undermining SSMs like the ADEA, by creating caste-ethnic divisions, a point that has already been alluded to in a previous quote.

> ADEA stand on a root called unity *(ekta)* and the promotion of unity will always be the primary requirement—a unity of minds, hearts and feeling of togetherness. The artificially created sense of difference, divisions and *jati-goshti* (caste-class feelings) need to be destroyed. Our *dhwaja* (flag) is unity (*ekta*) and we have to fly it high (*oraiba*). The flag that ADEA flies is of the people who have lost their land and their forests and who are losing their very roots . . . we must continue to create a political awareness around these issues, an adult education about society *(samajik shiksha)*. (ADEA Kondh *Adivasi* leader)

In addition to the subaltern caste-ethnic-class unity focus, micropolitical dimensions of significance include several other foci. Gender politics, as women's organizations grow in political maturity and confidence, oscillate between challenge and retreat in relation to male constructions of community life and gender relations, while apparently being consistently supportive of the broader politics of the ADEA in relation to globalist and developmentalist colonial challenges to the Adivasi way of life. As one Kondh woman leader remarks, "They (the Forest Department) ask us to create forest protection committees—protection from whom should I ask? . . . We need to protect the forest from them!" SSM-SSM coalitions and the politics of counterhegemonic possibilities must undergo various permutations in relation to inter-tribal and inter-caste unity. As tensions across generations pertaining to cultural change are created, the pace of change and the struggle in relation to processes of change by both youth and elders must engage movement learning and ADEA trajectories consistently. Finally, the local feudal and caste politics aimed at manipulating the ADEA in a localized game of competing interests along internal schisms defining ADEA subaltern groups, normally linked to macro-political forces of domination, must be addressed.

Learning Processes and Movement Learning in the ADEA

This movement learning dimension is focused on identifying (and answering) some of the following concerns. (a) Where does politically relevant learning take place (e.g., men often engage in political dialogue at the *salab basa* or local equivalent of a pub, while women engage in similar fashion at the village wells and other gender-specific meeting places). Such information has been used by the movement leadership to engage village communities more directly with ADEA concerns and to motivate and learn from village and social group-based perspectives of the same. (b) How has learning and different forms of knowledge been activated in the interests of the struggle on any number of fronts (e.g., legal knowledge such as the recent "right to information" stipulations and knowledge of land classification systems have been used successfully on numerous occasions to subvert attempts by forest guards and revenue officials in their attempts to threaten people off the land and out of the forests)? (c) How is Adivasi knowledge (cosmological, historical, ecological, agro-forestry) being used to massage movement learning, identity, motivation, strategy, and productive activity? What forms does this knowledge take (e.g., Elder narratives, songs, and lamentations)? (d) How does one participate in the politics of inside-outside knowledge (e.g., critiquing knowledge relations with NGOs who have often introduced new organizational technologies, like the village committee structure, that have subverted popular-democratic decision making or the Adivasi way of village discussion, sharing, and consensus building)? (e) How to access inside-inside knowledge and get involved in pedagogical engagements (between *Adivasis* and *Dalit* or between SSMs themselves). And, (f) How to develop content of organized teaching and learning for leadership gatherings and village-to-village or region-to-region gatherings or youth camps.

Catalytic Validity and Opportunities for Research-Instigated Movement Praxis

Research in subaltern movement contexts needs to (and will be rightfully interrogated in this respect by any seasoned movement organization engaged in a participatory research process) continually contribute toward movement learning, movement praxis, and the process of change being instigated through movement activism (Smith, 1992, 1999). For example, the research with the ADEA has helped to set up and establish the Center for Research and Development Solidarity (CRDS), an Adivasi-Dalit research and training organization which concerns itself with ADEA/community-specific knowledge processes; engaged a team of eight Adivasi-Dalit researchers in the research process who are consequently learning and shaping what it means to do social research in Adivasi contexts; and developed and floated a community/movement research sharing journal, *Amakatha* (Our Story), which has been actively used in the process of movement learning (the first of four issues dealt with Adivasi history, culture, and politics). The process of research learning in the ADEA movement has created numerous opportunities for movement introspective learning, enhanced movement identity through this self-interrogation, and actually helped movement leadership to rethink strategy and direction on a few occasions, leading to tangible outcomes in relation to land and resource control and Adivasi-Dalit unity within the movement.

Concluding Reflection: Globalization, Dispossession, and SSM Learning in the South

This chapter has provided a look at globalization and dependent-development/modernization as processes of political-economic and cultural imperialism (global and national capitalist colonizations) vis-à-vis subaltern social groups, has sketched the contours of SSMs and pointed to these formations in an attempt to disembed this social phenomenon from NSM/OSM taxonomies in the interests of a politics of the subaltern subject, and, finally, has elaborated on some emergent dimensions in relation to SSM learning in the South by drawing on research related to Adivasi SSMs in India. Taken together, it is hoped that a case has been made for the political and academic significance of SSMs as movement formations and as sites for critical and adult education and learning. In conclusion, subalterns are subjects of history and makers of their own destinies, despite encompassing structures of subordination. We need to continue to understand SSMs and their struggles in relation to the social orders, institutions, and the history of material-cultural capitalist colonial relations that mediate, shape, and influence the formation of subaltern subjects and their politics and learning.

Notes

1. A term first coined by Antonio Gramsci to refer to peasants. Used here rather loosely to refer to the above cited social groups (in the opening quote from Via Campesina) who are also defined by their sociocultural location as "ecological ethnicities" (Parajuli, 1990). Subalterity is also understood as the dialectics of superordination and subordination in global and national hierarchical social relations of exploitation.
2. The author acknowledges the assistance of the Social Sciences and Humanities Research Council (SSHRC) of Canada for my research in India through a Standard Research Grant (2006-2009). Reflections and grounded discussion in this chapter are both informed by this research.
3. See Giovani Arrighi's (1991; Arrighi et al., 2003) position pertaining to aggressive Northern neoliberal policies and the oligarchic wealth of the West as it draws capitalist activity toward it, widening the economic gap between regions (along a continuing North-South geographical axis) and subsequently entrenching a global hierarchy of nations of the world in the postcolonial era.
4. For examples, see Kwaipun's chapter on mining in Ghana in this collection; Broad, 2002; Davis, 2006; Paul & Steinbrecher, 2003; Peet & Watts, 2004; Robbins (2002); Shiva (2000); and Tucker (2007).
5. Such as the threat to use mass suicide by U'Wa indigenous peoples in Columbia if a proposed oil field on their territory were to be opened up (Cohen & Rai, 2000, p. 30) and similar such protests and struggles.
6. David Held (1995, pp. 163–172) suggests that indigenous peoples are burdened by "nautonomy" (a lack of autonomy) and structured disempowerment resulting from and in the asymmetrical production, distribution, and enjoyment of life chances. According to Cohen and Rai (2000, p. 19), "even within relatively benign hierarchies" (such as within GSM/NSM "progressive" coalitions) indigenous peoples experience a grossly nautonomic predicament; for example, the Sami in Scandinavia, First Nations in Canada, the Maori in New Zealand, and Aboriginal and Torres Strait Island people in Australia. According to Moody (1988), in Latin America, Asia, and Africa genocidal

conditions predominate. Choudry (2007) observes the same when it comes to pedagogies of mobilization in anticapitalist coalitions. See Kamat (2002) and Kapoor (2005) for NGO related depoliticizations of SSMs.

7. For useful discussions from indigenous standpoints, see Barker (2005) and Grande (2004).
8. See Holst's (2007, p. 8) proposition regarding socialist big utopias, objective communist movements, and the mundane day-to-day needs of populations on the margins of capital, or Hall's (2000) "blinding" (in the manner of omission of the very serious role of SSMs as historical and resilient agents of multiple politicizations in the face of hegemonic capitalist incursions that pre-date OSM/NSM/GSM categorization) conception of an all encompassing "global civil society" as the dominant protagonists of social change. Similar colonizations of the realm of a politics of protest and resistance include the apparent dwarfing of SSMs and their impacts through the evocative but nonetheless patronizing label "militant particularities." Social theorists of protest and change continue to exaggerate the importance of modernist hegemony and modern/postmodern forms of counterhegemony, while choosing to omit or failing to see the significance of continuous, multiple, day-to-day SSM politics being waged in the trenches of capital at considerable cost to its protagonists who harbor few illusions (given the daily benefit of continuous bouts with the fists of capitalist power/relations of power) concerning the power of the forces that they are compelled to address.

References

Abrahamsen, R. (2004). Review essay: Poverty reduction or adjustment by another name? *Review of African Political Economy, 99* (4), 187.

Allman, P. (1999). *Revolutionary social transformation: Democratic hopes, political possibilities and critical education.* Westport, CT: Bergin & Garvey.

Allman, P. (2001). *Critical education against global capitalism: Karl Marx and revolutionary critical education.* Westport, CT: Bergin & Garvey.

Alvarez, S., Escobar, A., & Dagnino, E. (Eds.). (1998). *Culture and politics, politics of cultures: Re-visiting Latin American social movements.* Boulder, CO: Westview Press.

Arrighi, G. (1991). World income inequalities and the future of socialism. *New Left Review, 189* (September-October), 39–66.

Arrighi, G., Silver, B., & Brewer, B. (2003). Industrial convergence, globalization and the persistence of the North-South divide. *Studies in Comparative Industrial Development, 38*(1), xx–xx.

Barker, J. (2005). *Sovereignty matters: Locations of contestation and possibility in indigenous struggles for self-determination.* Lincoln: University of Nebraska Press.

Behura, N., & Panigrahi, N. (2006). *Tribals and the Indian constitution: Functioning of the fifth schedule in Orissa.* New Delhi, India: Rawat Publications.

Bond, P. (2006). *Looting Africa: The economics of expoitation.* London: Zed.

Bowles, P. (2007). *Capitalism.* Harlow, UK: Pearson Education Ltd.

Broad, R. (Ed.). (2002). *Global backlash: Citizen initiatives for a just world economy.* New York: Rowman & Littlefield.
Cameron, A., & Palan, R. (2004). *The imagined economies of globalization.* London: Sage.
Carroll, W. (Ed.). (1997). *Organizing dissent: Contemporary social movements in theory and practice.* Toronto: Garamond Press.
Castells, M. (1997). *The information age: Economy, society and culture.* Vol. II. The power of identity. Oxford: Blackwell.
Chinweizu, M. (1975). *The west and the rest of us.* New York: Vintage.
Chossudovsky, M. (2003). *The globalization of poverty: Impacts of IMF and World Bank reforms.* New York: Zed.
Choudry, A. (2007). Transnational activist coalition politics and the de/colonization of pedagogies of mobilization: Learning from anti-neoliberal indigenous movement articulations. *International Education, 37*(1), 97–113.
Cohen, R., & Rai, S. (Eds.). (2000). *Global social movements.* New Bronswick, NJ: Athlone Press.
Cunningham, P. (1998). The social dimension of transformative learning. *PAACE Journal of Lifelong Learning, 7,* 15–28.
Cunningham, P. (2000). A sociology of adult education. In A. Wilson & E. Hayes (Eds.), *Handbook of adult and continuing education.* San Francisco: Jossey-Bass.
Davis, M. (2006). *Planet of the slums.* London: Verso.
Della Porta, D., & Diani, M. (1999). *Social movements: An introduction.* Oxford: Blackwell.
Escobar, A., & Alvarez, S. (Eds.). (1992). *The making of new social movements in Latin America: Identity, strategy and democracy.* Boulder, CO: Westview Press.
Fanon, F. (1963). *Wretched of the earth.* New York: Grove Press.
Fernandes, W. (2006). Development related displacement and tribal women. In G. Rath (Ed.), *Tribal development in India: The contemporary debate.* New Delhi, India: Sage.
Foley, G. (1999). *Learning in social action: A contribution to understanding informal education.* London: Zed.
Galeano, E. (1973). *The open veins of Latin America: Five centuries of the pillage of a continent.* New York: Monthly Review Press.
Grande, S. (2004). *Red pedagogy: Native American social and political thought.* Lanham, MD: Rowman & Littlefield.
Gray, J. (1997). *Endgames: Questions in late modern political thought.* Cambridge: Polity Press.
Hall, B. (2000). Global civil society: Theorizing a changing world. *Convergence, 23*(1-2), 10–32.
Held, D. (1995). *Democracy and the global order.* Cambridge: Polity Press.
Holst, J. (2002). *Social movements, civil society, and radical adult education.* Westport, CT: Bergin & Garvey.
Holst, J. (2007). The politics and economics of globalization and social change in radical adult education: A critical review of recent literature. *Journal for Critical Educational Policy Studies, 5*(1), 1–16.
Hoogvelt, A. (2001). *Globalization and the post-colonial world: The new political-economy of development* (2nd ed.). London: Palgrave.

Kamat, S. (2002). *Development hegemony: NGOs and the state in India.* Delhi: Oxford.

Kane, L. (2000). Popular education and the landless people's movement in Brazil (MST). *Studies in the Education of Adults, 32*(1), 36–50.

Kane, L. (2001). *Popular education and social change in Latin America.* London: Latin American Bureau.

Kapoor, D. (2005). NGO partnerships and the taming of the grassroots in rural India. *Development in Practice, 15*(2), 210–215.

Kapoor, D. (2007). Subaltern social movement learning and the decolonization of space in India. *International Education, 37*(1), 10–41.

Kapoor, D. (2008). Popular education and human rights: Prospects for antihegemonic Adivasi (original dweller) movements and counterhegemonic struggle in India. In A. Abdi & L. Shultz (Eds.), *Educating for human rights and global citizenship.* Albany, NY: SUNY.

La Belle, T. (1986). *Nonformal education in Latin America and the Caribbean—Stability, reform or revolution?* New York: Praeger.

Ludden, D. (Ed.). (2005). *Reading subaltern studies: Critical history, contested meaning and the globalization of South Asia.* New Delhi: Pauls Press.

Martinez-Alier, J. (2005). *The environmentalism of the poor: A study of ecological conflicts and valuation.* New Delhi: Oxford.

Mason, M. (1997). *Development and disorder: A history of the Third World since 1945.* Toronto: Between the Lines.

Mayo, P. (2004). *Liberating praxis: Paulo Freire's legacy for radical education and politics.* Westport, CT: Praeger.

McMichael, P. (2006). Reframing development: Global peasant movements and the new agrarian question. *Canadian Journal of Development Studies, 27*(4), 471–486.

McNally, D. (2002). *Another world is possible: Globalization and anti-capitalism.* Winnipeg, MB: Arbeiter Ring Publishing.

Mignolo, W. (2000). *Local histories/global designs: Coloniality, subaltern knowledges, and border thinking.* Princeton, NJ: Princeton University Press.

Moody, R. (1988). *The indigenous voice: Visions and realities, vol. 1 & 2.* London: Zed.

Moyo, S., & Yeras, P. (Eds.). (2005). *Reclaiming the land: The resurgence of rural movements in Africa, Asia and Latin America.* London: Zed.

Parajuli, P. (1990). *Grassroots movements and popular education in Jharkhand, India.* Ph.D. Diss., Stanford University.

Parel, A. (Ed.). (1997). *Hind Swaraj and other writings.* Cambridge: Cambridge University Press.

Paul, H., & Steinbrecher, R. (2003). *Hungry corporations: Transnational biotech companies colonize the food chain.* New York: Zed.

Peet, R., & Watts, M. (Eds.). (2004). *Liberation ecologies: Environment, development, social movements.* New York: Routledge.

Petras, J. & Veltmeyer, H. (2001). *Globalization unmasked: Imperialism in the 21st century.* Halifax, Canada: Fernwood Press.

Pimple, M., & Sethi, M. (2005). Occupation of land in India: Experiences and challenges. In S. Moyo & P. Yeras (Eds.), *Reclaiming land: The resurgence of rural movements in Africa, Asia and Latin America.* London: Zed.

Polet, F., & CETRI. (2004). *Globalizing resistance: The state of struggle.* London: Pluto.
Ray, R., & Katzenstein, M. (Eds). (2005). *Social movements in India: Poverty, power and politics.* Lanham, MD: Rowman & Littlefield.
Robbins, R. (2002). *Global problems and the culture of capitalism.* Boston, MA: Allyn and Bacon.
Rodney, W. (1972). *How Europe underdeveloped Africa.* London: Bogle-L'Ouverture.
Sheth, D. (2007). Micro-movements in India: Toward a new politics of participatory democracy. In D. Sousa Santos (Ed.), *Democratizing democracy: Beyond the liberal democratic canon.* London: Verso.
Shiva, V. (2000). The world on the edge. In W. Hutton and A. Giddens (Eds.), *Global capitalism.* New York: The New Press.
Sklair, L. (2001). *The transnational capitalist class.* Malden and Oxford: Blackwell.
Smith, G. (1992). *Research issues related to Maori education.* Auckland: University of Auckland.
Smith, L.T. (1999). *Decolonizing methodologies: Research and indigenous peoples.* London: Zed.
Sousa Santos, D. (Ed.). (2007). *Democratizing democracy: Beyond the liberal democratic canon.* London: Verso.
Starn, O. (1992). Peasant protest and the Rondas Campesinas of North Peru. In A. Escobar & S. Alvarez (Eds.), *The making of new social movements in Latin America: Identity, strategy and democracy.* Boulder, CO: Westview Press.
Tomlinson, J. (2001). Vicious and benign universalism. In F. Schuurman (Ed.), *Globalization and development studies: Challenges for the 21st century.* New Delhi, India: Vistaar Publications.
Torres, C. (1990). *The politics of nonformal education in Latin America.* New York: Praeger.
Tucker, R. (2007). *Insatiable appetite: The United States and the ecological degradation of the tropical world.* Lamham, MD: Rowman & Littlefield.
Via Campesina. (2005). Impact of the WTO on peasants in South East Asia and East Asia. Retrieved March 2, 2008, from http://www.viacampesina.org/en/images/stories//lvbooksonwto.pdf
Via Campesina. (2006). *Sovaranita alimentare, final declaration.* Retrieved March 1, 2008, from http://www.viacampesina.org/main_en/index.php?optioncom_content&task=view&id=180&Itemid=27.
Wignaraja, P. (1993). *New social movements in the South: Empowering the people.*

CHAPTER 6

Paulo Freire and Adult Education

Peter Mayo

Paulo Reglus Neves Freire (1921–1997) is surely one of the most cited and iconic figures in contemporary literature on education. In fact, his impact has been large enough to affect thinking in such areas as social work, communications, nursing, community development, theology, philosophy, and sociology. To provide one example, a recent text advocating a critical approach to community development (Ledwith, 2005) considers Freire's pedagogical and political philosophy as central to the strategy being outlined.

Freire spent his formative years in the city of Recife in Pernambuco State in Brazil, a city for which Freire always held tremendous affection. He spent the best part of his most creative years away from his homeland, forced into sixteen years of exile by the military regime that in 1964 overthrew the government led by the populist João Goulart. After having qualified as a lawyer, Freire decided to leave the profession and concentrate on education. He developed an approach to critical literacy involving "reading the word and the world," which captured the imagination of the government itself, especially its minister of education Paulo de Tarso Santos. Freire was asked to develop this program throughout the rest of Brazil. Freire must have been regarded as subversive by the ruling elite, notably members of the landowning classes and the bourgeoisie, if only because by helping many Brazilians become literate he was enabling them to vote. This by itself rendered the education imparted by Freire and his associates a political act. Add to this the fact that Freire's approach enabled people, through a process of praxis, to gain critical distance from their own reality and to perceive it in a more critical light, and one can begin to understand why the military regime, acting in the service of the local elites and foreign multinationals, would consider Freire a subversive educator engaged in "dangerous knowledge."

Freire's years in exile were spent briefly in Bolivia, Chile (this is where he and other Brazilian exiles gained greater exposure to Marxist ideas and where Freire developed the ideas for his celebrated work, *Pedagogy of the Oppressed*), the United States (*Pedagogy of the Oppressed* was published in English there), and Geneva, Switzerland. As a result of his work for the World Council of Churches in Geneva, Freire came into contact with people involved in liberation struggles in the former Portuguese colonies in Africa, as well as with different workers from Italy, Spain, Germany, and other parts of Europe, including immigrant "guest workers" (see Freire, 1994).

One of Freire's early works, *Education as the Practice of Freedom,* based on his doctoral dissertation, was rather liberal in tenor, unlike his more radical later works, starting from *Pedagogy of the Oppressed,* which draws on a variety of influences. It draws on Christian-Personalist theory (Mounier and Atiade), critical theory (Fromm, Marcuse), other Christian writings (Chardin and Niebuhr), anticolonial theory (Fanon, Memmi), and Marxist theory including Gramsci (to whom he was introduced to by Marcela Gajardo), Kolakowski, Kosik, and, noblesse oblige, Marx himself.

Freire drew on a wide range of early writings by Marx, notably, *The German Ideology,* The Economic and Philosophic Manuscripts of 1844, the Theses on Feuerbach, and *The Holy Family.* These writings provide important sources of reference for some of the arguments raised in Freire's best known work, *Pedagogy of the Oppressed* (1970b). Later writings by Marx, however, feature prominently in such works as *Pedagogy in Process* (Freire, 1978), where Freire attempts to come to grips with the social relations of production in an impoverished African country (Guinea Bissau) that had just gained independence from Portugal. In this work, and precisely in letter 11, Freire adopts Marx's notion of a "polytechnic education" (Castles & Wustenberg, 1979; Livingstone, 1984), arguing for a strong relationship to be forged between education and production (Freire, 1978). Marx had specifically developed this notion in the Geneva Resolution of 1866 (Livingstone, p. 187).

Most importantly, though, *Pedagogy of the Oppressed* is written in a dialectical style that, as Allman (1988) points out, is not easily accessible to readers schooled in conventional ways of thinking, often characterized by a linear approach. She demonstrates clearly that one cannot fully appreciate Freire's work without rooting it within Karl Marx's dialectical conceptualization of oppression. The more one is familiar with Marx's "tracking down" of "inner connections" and "relations" that are conceived of as "unities of opposites" (Allman, 1999, pp. 62, 63), the more one begins to appreciate Marxian underpinnings in *Pedagogy of the Oppressed* (see Allman, 2001, pp. 39–48). Among Freire's books, this is the most compact and consistent as far as the dialectical conceptualization of power is concerned (Allman et al, 1998).

Ideology

Freire's respective works are embedded in a Marxian conception of ideology based on the assumption that "the ruling ideas are nothing more than the ideal expression of the dominant material relationships, the dominant material relationships grasped as ideas; hence of the relationships which make one class the ruling one, therefore the ideas of its dominance" (Marx & Engels, 1970, p. 64). Not only does the ruling class produce the ruling ideas, in view of its control over the means of intellectual production (Marx & Engels), but the dominated classes also produce ideas that do not necessarily serve their interests; these classes, which "lack the means of mental production and are immersed in production relations which they do not control," tend to "reproduce ideas" that express the dominant material relationships (Larrain, 1983, p. 24).

Freire sees popular consciousness as being permeated by ideology. In his early works, Freire posited the existence of different levels of consciousness, ranging from naïve to critical consciousness, indicating a hierarchy that exposed him to the accusation of being elitist and of being patronizing toward ordinary people (Kane, 2001, p. 50). In these works, Freire reveals the power of ideology reflected in the fatalism (see Rossatto, 2005) apparent in the statements of peasants living in shantytowns who provide "magical explanations", attributing their poor plight to the "will of God" (Freire, 1970b, p. 163). A self-declared "man of faith," Freire extols the virtues of the "Prophetic Church," with its basis in liberation theology, and attributes "false consciousness" to the "traditionalist," "colonialist," and "missionary" church that he describes as a "necrophilia winner of souls" with its "emphasis on sin, hell-fire and eternal damnation" (Freire, 1985, p. 131).

Freire provides a very insightful analysis of the way human beings participate in their own oppression by internalizing the image of their oppressor. As with the complexity of hegemonic arrangements, underlined by Gramsci and elaborated on by a host of others writing from a neo-Gramscian perspective, people suffer a contradictory consciousness, being oppressors within one social hegemonic arrangement and oppressed within another. This puts paid to the now-hackneyed criticism that Freire's notion of oppressor and oppressed is so generic that it fails to take into account that one can be an oppressor in one context and oppressed in another. The notion of the oppressor and contradictory consciousness suggests otherwise. This consideration runs throughout Freire's oeuvre, ranging from his early discussion on the notion of the "oppressor consciousness" to his later writings on multiple and layered identities (Freire, 1997), where he insists that one's quest for life and for living critically is tantamount to an ongoing quest for the attainment of greater coherence. Gaining coherence, for Freire, necessitates one's gaining greater awareness of one's "unfinishedness" (Freire, 1998a, pp. 51, 66).

Resources of Hope

Freire accords an important role to agency in the context of revolutionary activity for social transformation. He explicitly repudiates evolutionary economic determinist theories of social change and regards them as being conducive to a "liberating fatalism" (Freire, 1985, p. 179), a position he adhered to until the very end, stating, at an honoris causa speech delivered at the Claremont Graduate University in 1989, that "when I think of history I think about possibility—that history is the time and space of possibility. Because of that, I reject a fatalistic or pessimistic understanding of history with a belief that what happens is what should happen" (Freire, in Darder, 2002, p. x).

The emphasis on voluntarism and on the cultural and spiritual basis of revolutionary activity is very strong in Freire's early writings, especially in the work based on his doctoral thesis. This particular aspect of his work is generally regarded to have been the product of strong Hegelian influences. The Hegelianism may have partly been derived from the writings of such Christian authors as Chardin, Mounier, and Niebuhr (Youngman, 1986, p. 159). In later writings, however, this idealist position becomes somewhat modified as Freire begins to place greater emphasis on the role of economic conditions in processes of social change.

Freire rejected the view that for the oppressed the conditions of their time determined the limits of what is possible. He recognized developments within capitalism, witnessed during his lifetime (neoliberalism), for what they were—manifestations of capitalist reorganization to counter the tendency of the rate of profit to fall, owing to the "crises of overproduction" (Allman & Wallis, 1995; Foley, 1999). Understanding the contemporary stages of capitalist development according to what they represented was a crucial step for Freire to avoid a sense of fatalism and keep alive the quest for attainment of a better world driven by what Henry A. Giroux (2001) calls an anticipatory utopia prefigured not only by a critique of the present but also by an alternative pedagogical/cultural politics. The fatalism of neoliberalism, buttressed by the propagation of an "ideology of ideological death" (Freire, 1998b, p. 14), was a key theme in Freire's later writings. It was the intended subject of the work he was contemplating at the time of his death (Araujo Freire, 1997, p. 10). Freire could well have been on the verge of embarking on an exploration of the conditions that the present historical conjuncture, characterized by neoliberalism, would allow for the pursuit of his dream of a different and better world. Alas, this was not to be.

Love

Freire was concerned with more than just the cognitive aspects of learning (Darder, 2002, p. 98). He regarded educators and learners as "integral human beings" (p. 94) in an educational process that has love at its core (p. 91). Just before he died, he was reported to have said: "I could never think of education without love and that is why I think I am an educator, first of all because I feel love."

The humanizing relationship between teacher and taught (teacher-student and student-teacher, in Freire's terms) is one characterized by love. It is love that drives the progressive Freire-inspired educator forward in teaching and working for the dismantling of dehumanizing structures. The entire process advocated by Freire is predicated on the trust he had in human beings and on his desire to help create "a world in which it will be easier to love" (Freire, 1970b, p. 24; see Allman et al., 1998, p. 9). This concept has strong Christian overtones as well as revolutionary ones. In the latter case, the influence could well have derived from Ernesto Che Guevara, who, according to Freire, "did not hesitate to recognize the capacity of love as an indispensable condition for authentic revolutionaries" (Freire, 1970a, p. 45).

Education in Its Broadest Context

The terrain for education action is a large one in Freire's conception. Throughout his writings, Freire constantly stressed that educators engage with the system and not avoid it for fear of co-optation (Escobar, Fernandez, & Guevara-Niebla, 1994; Horton & Freire, 1990). He exhorted educators and other cultural workers to "be tactically inside and strategically outside" the system. Freire believed that the system is not monolithic. Hegemonic arrangements are never complete and allow spaces for "swimming against the tide" or, to use Gramsci's phrase, engaging in "a war of position" (Freire, in Escobar et al., 1994, pp. 31, 32). In most of his works from the mid-1980s onward, Freire touches on the role of social movements as important vehicles for social change.

Freire himself belonged to a movement striving for a significant process of change within an important institution in Latin America and beyond, namely the Catholic Church. When he was education secretary in São Paulo, a position that allowed Freire to tackle education and cultural work in their broader contexts, he and his associates worked hard to bring social movements and state agencies together (O'Cadiz, 1995 O'Cadiz, Wong, & Torres, 1998;). These efforts on behalf of the Partido dos Trabalhadores (PT) continue to be exerted by the party itself in other municipalities, most notably in the city of Porto Alegre, in Rio Grande do Sul, where the PT had, until recently, been in government since the late 1980s, and presumably in the other municipalities and states where the party won the elections in the fall of 2000. There were also high hopes that these efforts would be carried out throughout the country once the PT leader Luiz Inacio "Lula" da Silva won the federal presidential elections, though perhaps too much was expected of Lula, who, in the words of many Brazilian sympathizers, won the government but not the state. The last years of Freire's life were exciting times for Brazilian society with the emergence of the Movimento dos Trabalhadores Rurais Sem Terra (MST), or Movement of Landless Peasants. The movement allies political activism and mobilization with important education and cultural work (see Kane, 2001, chap. 4.). The movement is itself conceived of as an "enormous school" (p. 97). As in the period that preceded the infamous 1964 coup, Paulo Freire's work and thinking must also have been influenced and reinvigorated by the growing movement for democratization of Brazilian society. In an interview with Carmel Borg and me, Ana Maria (Nita) Araujo Freire stated:

> Travelling all over this immense Brazil we saw and cooperated with a very large number of social movements of different sizes and natures, but who had (and continue to have) a point in common: the hope in their people's power of transformation. They are teachers—many of them are "lay": embroiderers, sisters, workers, fishermen, peasants, etc., scattered all over the country, in *favelas,* camps or houses, men and women with an incredible leadership strength, bound together in small and local organizations, but with such a latent potential that it filled us, Paulo and me, with hope

> for better days for our people. Many others participated in a more organized way in the MST (*Movimento dos Sem Terra*: Movement of Landless Peasants), the trade unions, CUT (*Central Única dos Trabalhadores*), and CEBs (Christian Base Communities). As the man of hope he always was, Paulo knew he would not remain alone. Millions of persons, excluded from the system, are struggling in this country, as they free themselves from oppression, to also liberate their oppressors. Paulo died a few days after the arrival of the MST March in Brasília. On that April day, standing in our living-room, seeing on the TV the crowds of men, women and children entering the capital in such an orderly and dignified way, full of emotion, he cried out: "That's it, Brazilian people, the country belongs to all of us! Let us build together a democratic country, just and happy!" (Nita Freire, in Borg & Mayo, 2000, p. 109)

Freire insisted that education should not be romanticized and that teachers ought to engage in a much larger public sphere (Freire, in Shor & Freire, 1987, p. 37). This has been quite a popular idea among radical activists in recent years, partly the result of a dissatisfaction with party politics. The arguments developed in these circles are often based on a very non-Gramscian use of the concept of "civil society." In his later works, however, Freire sought to explore the links between movements and the state (Freire, 1993; O'Cadiz et al., 1998) and, most significantly, between movements and party, a position no doubt influenced by his role as one of the founding members of the PT.

Freire argues that the party for change, committed to the subaltern, should allow itself to learn from and be transformed through contact with progressive social movements. One important proviso Freire makes in this respect is that the party should do this "without trying to take them over." Movements, Freire seems to be saying, cannot be subsumed by parties; otherwise, they lose their identity and forfeit their specific way of exerting pressure for change. Paulo Freire discusses possible links between party and movements: "Today, if the Workers' Party approaches the popular movements from which it was born, without trying to take them over, the party will grow; if it turns away from the popular movements, in my opinion, the party will wear down. Besides, those movements need to make their struggle politically viable" (Freire, in Escobar et al. 1994, p. 40).

Freire, therefore, explores the links between party and movements within the context of a strategy for social change. At the time when Paulo Freire was still alive, the PT enjoyed strong links with the trade union movements, the Pastoral Land Commission, the MST, and other movements and exercised a leadership role when forging alliances between party, state, and movements in the municipalities in which it was in power. The Participatory Budget project in Porto Alegre, an exercise in deliberative and participatory democracy, gives some indication of the direction such alliances can take (Schugurensky, 2002).

Praxis

The discussion has veered toward a macrolevel analysis. It would be opportune now to bring the discussion back to the microlevel of adult education with an emphasis on concepts that lie at the heart of the pedagogical relation as propounded by Freire. He regarded praxis as one of the key concepts in question. Praxis became a constant feature of his thinking and writing. It constitutes the means whereby one can move in the direction of confronting the contradiction of opposites in the dialectical relation (Allman, 1988, 1999). It constitutes the means of gaining critical distance from one's world of action to engage in reflection geared toward transformative action. The relationship between action, reflection, and transformative action is not sequential but dialectical (Allman, 1999). Freire and other intellectuals with whom he has conversed, in "talking books," conceive of different moments in their life as forms of praxis, of gaining critical distance from the context they know to perceive it in a more critical light. Exile is regarded by Freire and the Chilean Antonio Faundez (Freire & Faundez, 1989) as a form of praxis. The idea of critical distancing is however best captured by Freire in his pedagogical approach involving the use of codifications, even though one should not make a fetish out of this "method" (Aronowitz, 1993) because it is basically indicative of something larger, a philosophy of learning in which praxis is a central concept that has to be "reinvented" time and time again, depending on situation and context.

Authority and Freedom

Freire emphasized the notion of authentic dialogue throughout his works, regarding it as the means of reconciling the dialectic of opposites that characterizes the hierarchical and prescriptive form of communication he calls "banking education." Knowledge is not something possessed by the teacher and poured into the learner, who would thus be conceived of as an empty receptacle to be filled. This would be a static use of knowledge. Freire insisted on a dynamic process of knowledge acquisition based on epistemological curiosity involving both educator and educatee who regard the object of knowledge as a center of coinvestigation. Both are teachers and learners at the same time because teachers are prepared to relearn that which they think they already know through interaction with the learner, who can shed new light on the subject by virtue of insights including those conditioned by his or her specific cultural background. The learner has an important contribution to make to the discussion. Having said this, Freire warns against laissez-faire pedagogy that, in this day and age, would be promoted under the rubric of "learning facilitation." This is the sort of pedagogical treachery that provoked a critical response from Paulo Freire. In an exchange with Donaldo P. Macedo, Freire states categorically that he refutes the term *facilitator* (although he had used it earlier in such pieces as the essay in *Harvard Educational Review* concerning the literacy process in São Tome and Principe), which connotes such a pedagogy, underlining the fact that he has always insisted on the directive nature of education (Freire & Macedo, 1995, p. 394 Freire, in Shor & Freire, 1987, p. 103;). He insists on the term *teacher,* one who derives one's authority from one's competence in the matter being taught, without allowing this authority to degenerate into authoritarianism (Freire & Macedo, p. 378): "Authority is necessary to the freedom of the students and my own. The teacher is absolutely necessary. What is bad, what is not necessary, is authoritarianism, but not authority" (Freire, in Horton & Freire, 1990, p. 181; Freire, in Shor & Freire, 1987, p. 91).

Emphasis is being placed, in this context, on "authority and freedom," the distinction posed by Freire (see Gadotti, 1996), who argues that a balance ought to be struck between the two elements. In *Pedagogy of Hope,* Freire argues that the educator's "directivity" should not interfere with the "creative, formulative, investigative capacity of the educand." Otherwise, the directivity degenerates into "manipulation, into authoritarianism" (Freire, 1994, p. 79). Referring to this aspect of Freire's work, Stanley Aronowitz is on target when he states, "The educator's task is to encourage human agency, not mold it in the manner of Pygmalion" (1998, p. 10). The encouragement of human agency is a key feature of the work of Paulo Freire.

Some Criticisms

Needless to say, Freire has had his critics over the years. Some have argued that his vision is anthropocentric, a fair comment on Freire's works, especially his earlier works, although it has to be said that the institute he helped give rise to, now the Paulo Freire Institute, is working hard within the context of the Earth Charter in the area of ecopedagogy. He has also been the target of several attacks by feminists concerning what bell hooks (1993, p. 148) regards as his "phallocentric paradigm of liberation," although hooks would always affirm the validity of Freire's works in a process of liberation, and she draws extensively from them. Her book *Talking Back* (1989) is full of citations from *Education for Critical Consciousness* and *Pedagogy of the Oppressed*. Recent works that combine Freire's insights with those deriving from feminist theory and practice for a "Freirean feminist praxis" of community development include Margaret Ledwith's (2005) text on community development. Collaborators from North America, especially Donaldo Macedo, also pressed Freire hard, in interviews, on the gender issue. It is to be said that, in numerous later works, Freire tackled the issues of machismo and gender, as well as issues concerning multiple and contradictory subjectivities (Freire, 1997). Of course, despite his preference for talking books in the late 1980s and early 1990s, we never came across any such book in English dealing with either a woman or a person of color that would have made discussions concerning patriarchy and "race" sustained. One major criticism of these talking books is that they broach many subjects but not always in the depth required. The dialogical book format conditions both speakers to flit from one topic to another.

Others see contradictions in the fact that, despite Freire's emphasis on dialogue, it is always the teacher who holds the cards (Coben, 1998) to which the distinction between authority and authoritarianism, mentioned earlier, and the dangers of laissez-faire pedagogy, are referred in defense of what is a complex notion of dialogue. Of course, unless the educators are well prepared, there is always the danger of having a travesty of Freirean pedagogy (see Bartlett, 2005 for a discussion on the limits and possibilities of Freirean pedagogy; see also Stromquist, 1997). One standard criticism of Freire's works is his supposed use of "binary opposites" (e.g., oppressor-oppressed, subject-object, action-reflection) when in actual fact the relations, as in Marx, are dialectical (Allman, 1999) and therefore the two elements are intimately connected (Darder, 2005, p. 92).

Other criticisms concern Freire's overemphasis on popular culture; he eschews discussions concerning high-order thinking and high-status culture in his works, which contrasts with the works of, say, Gramsci, who places emphasis on the critical appropriation of the dominant culture (Mayo, 1999). There is little such material concerning high-order thinking and high-status culture even with regard to the interdisciplinary curriculum, based on generative themes, of the "popular public" schools in São Paulo, even though the reform did not last long enough for anyone to be able to witness the transition from the popular and situationally embedded knowledge to the high-order thinking and knowledge that can prevent pupils from remaining on the periphery of political life. This issue concerns not only the popular public school, which benefited from insights derived from popular education, but also popular and adult education.

Finally, one major criticism of Freire is that, in his later years, he produced one book too many, especially as far as books in the English language are concerned, and there is quite a lot of repetition across these books. On the other hand, there are those who would lament that some excellent talking books, such as the exchange with Frei Betto and Ricardo Kotscho (Betto & Freire, 1986) and with Moacir Gadotti and Sergio Guimarães (Gadotti, Freire, & Guimarães, 1995), never saw the light in English translation.

Despite these criticisms, Paulo Freire stands out as one of the towering figures of twentieth-century educational thought. He has provided us with a huge corpus of literature containing ideas that can inspire people committed to the fostering of greater social justice. It is now left to others to make creative use of his theoretical and biographical legacy with a view to making sense of the contexts in which they operate, through a process of reinvention and not transplantation.

References

Allman, P. (1988). Freire, Gramsci and Illich. Their contribution to radical education for socialism. In T. Lovett (Ed.), *Radical adult education: A critical reader* (pp. 58–113). London: Routledge.

Allman, P. (1999). *Revolutionary social transformation: Democratic hopes, political possibilities and critical education.* Westport, CT: Bergin & Garvey.

Allman, P. (2001). *Critical education against global capitalism: Karl Marx and revolutionary critical education.* Westport, CT,: Bergin & Garvey.

Allman, P. (with Mayo, P., Cavanagh, C., Lean Heng, C., & Haddad, S.). (1998). The creation of a world in which it will be easier to love. *Convergence,* 31 (1 & 2), 9–16.

Allman, P., & Wallis, J. (1995). Challenging the postmodern condition: Radical adult education for critical intelligence. In M. Mayo & J. Thompson (Ed.), *Adult learning critical intelligence and social change* (pp. 18–33). Leicester, UK: NIACE.

Araujo Freire, A.M. (1997). A bit of my life with Paulo Freire. *Taboo: The Journal of Culture and Education,* 11 (Fall), 3–11.

Aronowitz, S. (1993). Freire's radical democratic humanism. In P. McLaren & P. Leonard (Eds.), *Paulo Freire: A critical encounter* (pp. 8–24). London: Routledge.

Aronowitz, S. (1998). Introduction. In Freire, *Pedagogy of freedom.* Lanham, MD: Rowman & Littlefield.

Bartlett, L. (2005). Dialogue, knowledge and teacher-student relations. Freirean pedagogy in theory and practice. *Comparative Education Review,* 49 (3), 344–364.
Betto, F., & Freire, P. (1986). *La scuola chiamata vita.* Bologna: EMI.
Borg, C., & Mayo, P. (2000). Reflections from a "third age" marriage: Paulo Freire's pedagogy of reason, hope and passion. An interview with Ana Maria (Nita) Araujo Freire. *McGill Journal of Education, 35* (2), 105—120.
Castles, S., & Wustenberg, W. (1979). *The education of the future: An introduction to the theory and practice of socialist education.* London: Pluto.
Coben, D. (1998). *Radical heroes: Gramsci, Freire and the politics of adult education.* New York: Garland.
Darder, A. (2002). *Reinventing Paulo Freire: A pedagogy of love.* Boulder, CO: Westview Press.
Darder, A. (2005). What is critical pedagogy? In W. Hare & J.P. Portelli (Eds.), *Key questions for educators.* Halifax: EdPhil Books.
Escobar, M., Fernandez, A.L., & Guevara-Niebla, G. (with Freire, P.). (1994). *Paulo Freire on higher education: A dialogue at the National University of Mexico.* Albany: SUNY Press.
Foley, G. (1999). *Learning in social action: A contribution to understanding informal education.* London: Zed Books.
Freire, P. (1959). *Education in present-day Brazil* PhD these completed at the University of Recipe, Brazil.
Freire, P. (1970a). *Cultural action for freedom.* Cambridge, MA: Harvard University Press.
Freire, P. (1970b). *Pedagogy of the oppressed.* New York: The Seabury Press.
Freire, P. (1973). *Education for critical consciousness.* New York: Continuum.
Freire, P. (1978). *Pedagogy in process: The letters to Guinea Bissau.* New York: Continuum.
Freire, P. (1985). *The politics of education.* South Hadley, MA: Bergin & Garvey.
Freire, P. (1993). *Pedagogy of the city.* New York: Continuum.
Freire, P. (1994). *Pedagogy of hope.* New York: Continuum.
Freire, P. (1997). A response. In P. Freire (Ed.). (with J.W. Fraser, D. Macedo, T. McKinnon, & W.T. Stokes), *Mentoring the mentor: A critical dialogue with Paulo Freire.* New York: Peter Lang.
Freire, P. (1998a). *Pedagogy of freedom: Ethics, democracy and civic courage.* Lanham, MD: Rowman & Littlefield.
Freire, P. (1998b). *Teachers as cultural workers: Letters to those who dare teach.* Boulder, CO: Westview Press.
Freire, P., & Faundez, A. (1989). *Learning to question: A pedagogy of liberation.* Geneva: World Council of Churches.
Freire, P., & Macedo, D. (1995). A dialogue: Culture, language and race. *Harvard Educational Review, 65* (3), 377–402.
Gadotti, M. (1996). *Pedagogy of praxis: A dialectical philosophy of education.* Albany: SUNY Press.
Gadotti, M., Freire, P., & Guimarães, S. (1995). *Pedagogia: Dialogo e conflitto.* (B. Bellanova & F. Telleri, Eds.). Torino: Societa' Editrice Internazionale.

Giroux, H. (2001). *Public spaces/private lives: Beyond the culture of cynicism.* Lanham, MD: Rowman & Littlefield.

hooks, b. (1989). *Talking back: Thinking feminist, thinking black.* Toronto, ON: Between the Lines.

hooks, b. (1993). bell hooks speaking about Paulo Freire. The man, his works. In P. McLaren & P. Leonard (Eds.), *Paulo Freire: A critical encounter* (pp. 146–154). New York: Routledge.

Horton, M., & Freire, P. (1990). *We make the road by walking: Conversations on education and social change.* Philadelphia: Temple University Press.

Kane, L. (2001). *Popular education and social change in Latin America.* London: Latin American Bureau.

Larrain, J. (1983). *Marxism and ideology.* Atlantic Highlands, NJ: Humanities Press.

Ledwith, M. (2005). *Community development: A critical approach.* Bristol: BASW/Policy Press.

Livingstone, D.W. (1984). *Class, ideologies and educational futures.* Sussex: The Falmer Press.

Marx, K., & Engels, F. (1970). *The German ideology.* (C.J. Arthur, Ed.). London: Lawrence and Wishart.

Mayo, P. (1999). *Gramsci, Freire and adult education: Possibilities for transformative action.* London: Zed Books.

O'Cadiz, M. (1995). Social movements and literacy training in Brazil: A narrative. In C.A. Torres (Ed.), *Education and social change in Latin America.* Melbourne: James Nicholas.

O'Cadiz, M., Wong, P.L., & Torres, C.A. (1998). *Education and democracy: Paulo Freire, social movements and educational reform in São Paulo.* Boulder, CO: Westview Press.

Rossatto, C.A. (2005). *Engaging Paulo Freire's pedagogy of possibility: From blind to transformative optimism.* Lanham, MD: Rowman & Littlefield.

Schugurensky, D. (2002). Transformative learning and transformative politics: The pedagogical dimension of participatory democracy and social action. In E., O'Sullivan, A. Morrell, & M. O'Connor (Eds.), *Expanding the boundaries of transformative learning.* New York: Palgrave.

Shor, I., & Freire, P. (1987). *A pedagogy for liberation: Dialogues on transforming education.* South Hadley MA: Bergin & Garvey.

Stromquist, N. (1997). *Literacy for citizenship; Gender and grassroots dynamics in Brazil.* Albany: SUNY Press.

Youngman, F. (1986). *Adult education and socialist pedagogy.* Kent: Croom Helm.

CHAPTER 7

Freire and Popular Education in Indonesia: Indonesian Society for Social Transformation (INSIST) and the Indonesian Volunteers for Social Transformation (Involvement) Program

Muhammad Agus Nuryatno

This chapter discusses the experience of the Indonesian Society for Social Transformation (INSIST),[1] a nongovernmental organization (NGO) based in Jogjakarta, in conducting the educational program known as Indonesian Volunteers for Social Transformation (Involvement), a transformative course for NGO activists, from 1999 to 2003, funded by KEPA, Finland.[2] The training programs offered by INSIST are based on Freirean critical pedagogy, on account of the possibility that neglecting this approach would contribute to the degeneration of critical capacities among NGO activists.

After a brief overview of development issues in Indonesia today and the subsequent rationale for the establishment of INSIST and programs like Involvement, this chapter describes the Involvement program in terms of its participants, pedagogical emphasis, and its impact on social change. Finally, critical reflections pertaining to this program are shared in the interest of enhancing its contribution toward social change processes that address the interests of marginalized groups in Indonesia.

Indonesian Development and the Rationale for "Involvement"

With the downfall of the Suharto regime in 1998, Indonesia entered a transitional phase as it passed from an authoritarian to a democratic government. Instituting a democratic society has not been an easy task because Indonesia was for so long governed by an authoritarian regime that emphasized cultural uniformity, political stability, and social order at the expense of democratic and egalitarian values (A. A. Nugroho, personal communication, June 11, 2003). To pave the way for a democratic Indonesia in which the politics of difference is accommodated, it is necessary to strengthen the components of civil society and particularly NGOs, which apparently have not yet found the best way to undergo this transitional process (M. Fakih, personal communication, July 28, 2003).

The Indonesian government today is dominated by neoliberal views, that is, a political-economic philosophy that de-emphasizes or rejects government intervention in the domestic economy—one impact of its position as a client of the International Monetary Fund (IMF) (1997–2002). The implementation of neoliberalism in Indonesia actually began in the mid-1980s with the introduction of deregulation and debureaucratization (Djamhari, 1996). However, its full implementation gained momentum after Indonesia was hit by a monetary crisis in mid-1997, at which point the Indonesian government formally asked the IMF to help in the economic recovery.

As a consequence, via signing a letter of intent (LOI), Indonesia has aligned itself with the Washington Consensus (*The Jakarta Post,* 1997). One of its requirements is to eliminate subsidies for refined fuel oil, thus paving the way for multinational companies, such as Shell, to operate in Indonesia. Another point in the agreement is to privatize state-owned companies, such as Indosat and Telkom (telecommunications), the National Bank of Indonesia, Tambang Timah, and Aneka Tambang. Worse, water is opened and directed to privatization through Water Law No. 7/2004, which is no doubt in line with the World Bank's scheme for water resources sector adjustment loans (Hadad, 2003). Thus, water will no longer be free to consume or publicly supported; it will be treated as a commodity on the open market. People will have to buy it from corporations, either state/private national corporations or multinational/transnational corporations.

Not only some state-owned corporations but also state universities have been privatized. Education Law No. 61/1999 effected the change in status of selected state universities, such as the Indonesian University,

Gadjahmada University, Bandung Institute of Technology, and Bogor Institute of Agriculture, into BHMN (Badan Hukum Milik Negara—state-owned law board). This change in status gave a kind of legitimacy to the privatization of public education because, with BHMN status, universities gained the autonomy needed to seek sources of funding other than government subsidies. They have the autonomy and independence to increase tuition fees and create programs that can generate money for the university. For example, they have created a program for tracing special interests and competency, known as the "special lane," a new model of recruiting students that complements the system of accepting new students. The latter model is purely based on academic standards. The former, however, makes wealth the primary consideration in accepting new students. The more a student is able to pay to the university the more likely he/she is to be accepted.

This ideology of neoliberalism has now become widespread in Indonesia and has affected many social institutions including the state bureaucracy, schools, and NGOs. Surprisingly, many NGOs unconsciously support the IMF agenda, albeit indirectly. This can be seen in their programs that insist on good governance, civil society, and land reform—three issues that have become conditional for NGOs that want to obtain foreign funding. However, it is impossible to get funding if a proposal betrays any intention to monitor international companies operating in Indonesia (Fakih, personal communication, 2003). The hidden agenda is to weaken state institutions and to protect transnational companies from potential criticisms of NGOs.

Critics of these developments insist that, in order to counter the hegemony of neoliberalism and pave the way for a democratic Indonesia in which the politics of difference is accommodated and human dignity is respected, the components of civil society and particularly the NGOs, need to be strengthened. Indonesian NGOs share many core concepts and values, among them, self-reliance, participation, and democracy. However, they differ in terms of the ideologies they propound and their approaches to certain questions, including differences in levels of militancy. Philip J. Eldridge (1995), who studied Indonesian NGOs from 1983 to 1993, established four paradigms in relation to the regime. Although the study was established to portray NGOs under the New Order regime, it can also be applied in the context of post–New Order regimes.

The first group can be said to include organizations dedicated to 'high-level co-operation grass-roots development.' NGOs in this category are consistent in promoting core values such as self-reliance and participation, but they are not interested in the political transformation of the regime. Their decision to adopt a nonpolitical approach is to protect their autonomy from government interference. Members of the second group can be said to aim at "high-level politics grassroots mobilization." Although NGOs in this category were critical of the New Order philosophy and initially refused to participate in government-sponsored development programs, they eventually chose to cooperate with the government—and continue to do so—for projects such as the development of water supply in urban sectors and environmental management programs.

The organizations in the third group focus more on building awareness among the local people than on efforts to change state policy. Organizations of this type can be described as favoring "empowerment from below" because they pay more attention to small communities and avoid getting involved in large-scale networking arrangements. The three groups mentioned thus far have been criticized by the fourth group for lacking militancy and for not mobilizing the grass roots against the regime due to their conflict-avoiding strategy and developmental ideology. They are accused of having failed to propose an alternative model of development and of neglecting to form an opposition movement dedicated to representing the poor. As a result, according to the fourth group, they have failed to accommodate the interests of workers and peasants and to shelter them from the domination of global capitalism represented by multinational companies, and a more radical movement for structural transformation is therefore necessary.

The failure of most Indonesian NGOs to act as agents of social change is affirmed by Mansour Fakih (1996), who argues that Indonesian NGOs have not been able to find real solutions for empowering people due to their ideological preference for developmentalism. Contradictions are inherent in many big NGOs when it comes to their jargon or theory and actual practices. For example, they all claim democracy, social transformation, and justice as their core values and concepts, yet employ modernization and developmentalist theories to achieve such goals from a noncritical standpoint. This theoretical ambiguity leads to inconsistencies in methodology and approach, resulting in the co-optation and domestication of radical and critical methodologies and techniques, such as Freirean conscientization education and participatory action research.

To respond to the above crisis, a group of NGO activists held a series of meetings starting in the late 1980s to find a new definition for the role of NGOs in Indonesia. The ultimate goal of the meetings was to recapture the original vision and mission of the NGOs as agents of social change. The meetings took place in Bukittinggi, North Sumatra in 1987; Baturaden, central Java, in 1990; and Cisarua Bogor, East Java, in 1993 (Fakih, 1996). In these meetings, some critical reflections were voiced: (1) establishing democratic and just internal structures in NGOs; (2) insisting on advocacy and people's organizations instead of short-term projects that stifle popular struggle; (3) taking critical account of the hegemonic structure of international capitalism disseminated through foreign funding, such as USAID, CIDA, and the World Bank; and (4) changing the vision of NGOs from a traditional and reformist mode to a transformative paradigm.

One of the suggestions on how to achieve the above was to reschool NGO activists in order to equip them with the analytical and theoretical competencies necessary to locate NGOs in relation to the state and the hegemony of developmentalist ideology. It was in this context that Involvement was significant. The school provides the activists with not only basic competencies such as analytical skills, but also emancipatory knowledge such as social movement theory, participatory action research, critical theory, and popular education. These ideas are grounded in the works of Paulo Freire, Saul David Alinsky, Antonio Gramsci, and other social theorists and critical pedagogues.

"Involvement"

The participants in Involvement are usually student or NGO-based activists or other social activists who have at least one year's experience in addressing civil society issues and who are committed to work for the empowerment of civil society (*Laporan Kegiatan Kelas,* 2000). The participants should have acquired experience in a social movement organization aimed at empowering, for instance, workers, women, peasants, fishermen, and indigenous communities. According to Noer Fauzi (quoted in Sangkoyo, 2003), the program is designed for talented activists, the educated middle class, or the "petit bourgeois" type of activist.

There are two ways of recruiting participants. First, INSIST, as organizing committee of "Involvement," asks local NGOs to send their members to participate in this program. Second, INSIST chooses the participants based on the recommendations of its senior members (*Laporan Kegiatan Kelas,* 1999). Participants in the training can be divided into two types: those who are well informed about social theories but lack fieldwork experience, and those who are rich in fieldwork experience but lack theoretical knowledge. It is difficult to find participants who combine both aspects, for they usually possess one and not the other.

Since 1999, more than one hundred activists have graduated from the Involvement program and are now active in many places throughout Indonesia, including Java, Sumatra, Maluku, Bali, East Timor, Papua, and Kalimantan. Topatimasang (2002, in Sangkoyo, 2003) maintains that the alumni of Involvement are oriented not toward establishing new organizations or NGOs, but rather toward supporting local initiatives in people's organizing, arguing that many activists are trapped in the dominant trend of establishing new organizations in order to get foreign funding instead of strengthening existing people's institutions and organizations. Thus, after graduating, the participants are expected to take part as volunteers in any social organization, including that of peasants, workers, or indigenous communities.

Pedagogy: Formal Curricular Engagements

There are three stages in the educational program of Involvement: two months of classroom work, nine months of fieldwork, and one month of evaluation. Fauzi (Sangkoyo, 2003) points out that "the two-month session is especially instrumental in shaping their political outlook" (p. 22). The dynamics of learning vary from one class to another, depending on two important factors: the participants and the facilitators. The process of learning in Involvement is based on a Freirean approach, in the sense that it is the participants who decide what kind of knowledge they want to gain, how to achieve it, and who will function as facilitators in the class—assuming that they know what they need. The learning process is thus organized from the bottom rather than from the top of the pyramid. However, the committee organizer provides alternative subjects that can be adopted by participants as guidelines in formulating syllabi.

In terms of facilitators, the participants prefer people who not only have a good knowledge of the subject but also have experience in the field. For example, Roem Topatimasang, who has extensive experience in organizing groups of people throughout Indonesia (Maluku, Papua, Aceh, East Timor, and others), always presents on community organization to most classes. The materials taught in the two-month session include, among others, (1) schools of thought in education, (2) popular education, (3) theories of social transformation, (4) social analysis theory and community organization, and (5) research methodology.

The schools of thought in education involve the study of three such schools, conservative, liberal, and transformative-critical. These schools have their own preferences in terms of theoretical perspective, methodological framework, content of learning, ends of learning, and methods.

The Involvement program is consciously based on a transformative-critical approach that is inspired by critical pedagogues such as Paulo Freire because the process of learning is ultimately oriented toward increasing the critical capacities of the participants so they may understand how the social system works.

The main focus of popular education is Paulo Freire because he is seen as the main exponent and source of ideas. Two of his works (in their Indonesian versions) are discussed intensively, *Pedagogy of the Oppressed* (1971) and *The Politics of Education: Culture, Power, and Liberation* (1985). This subject is usually delivered by Mansour Fakih, not only because he has mastered the subject but also because he met with Freire in 1989 when pursuing his Ph.D. at the Massachusetts Institute of Technology. At this point, particular attention is given to Freire's archaeological concept of consciousness, his notion of education as politics, and his vision of literacy.

Theories of social transformation is another subject taught in Involvement classes. This subject is important for NGO activists because, as actors in social movements, they need to know how social transformation works. It is argued that social changes at the national and local levels cannot be separated from global transformation. For this reason, international events, such as the two world wars and the establishment of the United Nations, the World Bank, the IMF, the Marshal Plan, the Colombo Plan, and the like, are discussed and analyzed critically in relation to their impact on local issues in Indonesia. The shift in models of capitalism from liberal to state capitalism, to neoliberal, is also analyzed extensively by the participants.

Another subject taught in Involvement pertains to community organization. The facilitator selected to present this subject is usually Roem Topatimasang, who has devoted himself to organizing communities throughout Indonesia. Topatimasang argues that community organizing is oriented to help people become aware, to know, and be able to act. Decision making is thus in the hands of the people, not the activists. Based on his experiences, Topatimasang (2000) says that there are three principles in organizing a community. First, it is impossible to develop people's organization without becoming directly involved. Second, it is important and necessary to be well informed about the people in the community being organized. Third, it is important not to depend too much on foreign funding. In the early stages foreign funding may be necessary, but in the long run it can become a burden. Therefore, it is important to develop self-reliance in terms of economic resources in order to remain independent.

The issue of foreign funding is crucial in community organizing because it is a double-edged sword. On the one hand, it can be used to run programs designed to empower the people. On the other, it can have the effect of weakening the social system, in the sense it causes people to lose their creativity and abandon their culture of *gotong-royong* (mutual assistance), which has been embedded within local communities for centuries. People as a result become dependent beings and always look to outsiders to solve the problems arising in their villages. When an NGO or other social institution comes to the village, what is uppermost in the minds of people is that these institutions will bring money. Surprisingly, many social institutions perpetuate this kind of dependency and make less of an effort to educate people to be self-reliant.

The subjects covered in research methodology focus on the different perspectives held by the positivist and nonpositivist approaches. The positivist approach is based on the assumption that humans are part of the natural world, which is orderly, predictable, and knowable, and that their behavior can also be understood by looking at the laws of cause and effect. Human behavior and social actions are therefore kinds of natural events that occur outside the scope of human consciousness. The positivist approach has been criticized for its claim to represent value-free research, which is considered an unattainable goal. This paradigm is also accused of supporting the existing social order in the name of neutrality and nonbiased research.

The basic assumption of the nonpositivist approach is that research should be conducted to transform the living conditions of the people being studied. For this purpose, instead of proposing internal (coherence) and external (isomorphism) validity by assuming that true knowledge is a mirror of reality, this approach proposes catalytic validity, whereby it strives to ensure that research leads to action and that it transforms the living conditions of the participants. The most remarkable difference in the nonpositivist approach, particularly in the case of critical inquiry, is that it is praxis oriented. There is a strong belief that theory, when allied with praxis, leads to a proper political end, namely, social transformation. The goal of research is, therefore, to critique, emancipate, and improve the condition of humanity. Thus, research should have a kind of "transformative agenda."

Method of Learning

Pedagogical process in Involvement is based on the dialogical method. Topatimasang, for example, often begins a session on community organizing by asking the participants to watch a fifteen-minute video about the life of a community in the village of Aru, Maluku, which is very poor despite the presence of vast amounts of pearls, a very expensive and promising international commodity. The participants are then asked to discuss and analyze the problems of this society. The first question to be investigated is, Who is involved in this social problem? After the problems have been defined and the actors identified, the subject is addressed.

In one dialogue, Dahlan Tamher (2001), one of the participants in the 2000 class, posed a challenging question to Topatimasang: "What should we do if we find out that a decision made by the *adat* (customary) institution disadvantages the local people, for if we want to challenge it we will be isolated from the rest of the community?"

In response to this question, Topatimasang insisted that the solution in this case is creativity and methodology. He gave an example from his own experience. In Maluku, there is a village where the social structure is very patrilineal. Women are not allowed to attend the *adat* assembly. Topatimasang did not address this gender issue right away when he approached the community, because he knew that had he done so, he would have been isolated and would not have been accepted by the local people. Instead, he got involved in people's activities for four years until they trusted him, and only then did he ask the question, "Why are women not allowed to attend the *adat* meeting?" This gender issue was then discussed by influential local figures (like the heads of the *adat*), and it appeared that such regulations were not really part of customary law. It was even pointed out that women had in the past contributed to the formation of customary law. Two years later, women were allowed to participate in the *adat* assembly.

Pedagogy: Praxis in the Communities

At the end of their two months in the classroom, the participants are asked to plan their nine-month fieldwork, particularly in terms of where they want to go, which institutions they want to join, and the role they are willing to take. In general, most of the participants choose their own communities as the setting for their fieldwork. Most of the participants of the 1999 class, for example, went back to their original institutions. Juliana Jamlean, Ahmad Syakir Renwarin, Muhammad Mashuri Kabalmay, and Efrem Silubun returned to their home institutions, Yayasan Nen Mas Il, Tual Maluku Tenggara (Kei Kecil), and Yayasan Pengembangan Maur Ohio Wut Maluku Tenggara (Kei Besar), respectively. There they worked at organizing the indigenous communities and helping the local people to develop their economy. Likewise, Putut Indrianto returned to his original institution, Keliling, Klaten, to counsel peasants in the local community. Others chose locations that were basically new and different from their original institutions, such as Biduran Kaplale and A. M. Indrianto. Instead of returning to Baileo Maluku, Kaplale did his fieldwork in Mitra Tani, Jogjakarta. Likewise, instead of returning to the E-Team in Jogjakarta, Indrianto joined Yayasan Lingkungan Hidup (YALI) Jayapura, West Papua, to do his fieldwork.

Doni Hendro Cahyono (personal communication, July 29, 2003), an alumnus of the 1999 Involvement class, did his fieldwork in Magelang, central Java. He tried to apply some of the precepts he had learned during his two months in class when organizing the workers in this area, and had considerable success. Women workers, who were used to sitting in the back quietly during discussion forums, wanted to participate more actively. Since a bond of trust between the workers and the community organizer had been established, the workers did not hesitate to communicate their domestic problems directly to Cahyono, realizing that individual problems could possibly disturb the collective action. The community organizer always exploits informal meetings as an entry point to strengthening solidarity amongst workers.

At this point, Cahyono employed Freirean principles to organize workers by defining them not as objects but as subjects who should become the main actors in solving their own problems. He argues that a true community organizer is not one who presents himself as a hero come to save the people; the role of the community organizer is only to encourage and help the community become self-reliant. He/she does not have the right to become the de facto leader of that community. In fact, the leader should be from the community itself. Any union, whether of peasants, workers, or fishermen, should be led by someone who is genuinely part of that community; thus one who has no experience as a worker is ethically ineligible to lead a workers' union. These principles should be taken into account by every community organizer. Otherwise the process of organizing a community may result in dependence instead of self-reliance.

Solihul Hadi (personal communication, July 29, 2003) tells a different story concerning his experience in conducting fieldwork with the Bajo tribe of East Kalimantan, which is a fishing community. The Bajo tribe is distributed between two islands: Derawan, which has one village, and Maratua, which has four. The trip from Derawan to Maratua takes four hours in a fishing boat, and to go from one village to another on Maratua takes two hours by foot. This tribe was previously unknown to Hadi, and none of his friends had undergone a similar experience with a tribe. Thus, he could not take his colleagues' experiences as a model. How did Hadi deal with this situation?

First of all, he recalled the teaching he had received during his two-month class and tried to contextualize the theories and adapt them to the present situation, particularly in regard to organizing the community and popular education. Hadi took the core ideas of the subjects he had discussed in class and tried to ground them in reality. This is the most difficult part for any social activist. For instance, the main task of a community organizer is to empower the local people and help them become self-reliant. For Hadi, it meant that in politics people should be involved in deciding public policies, in economic matters they should dominate the market in order to be able to fix the prices of commodities and of what they produce, and at the social-cultural level they have to learn how to be proud of their own culture. The essence of community organizing and popular education, according to Hadi, is to define the local inhabitants as subjects in the process of social transformation. Methods of empowerment are based on "what they know, what they have, and what they can do."

The first thing to learn is how to become part of the local community and how not to pass for a stranger. Hadi tried to become a friend of the villagers. Being accepted as part of the community is an essential step toward successful community organizing, because it is only after having been accepted that one can identify what the problem is, who is involved in the problem, and what the relations among the actors are. The main problem faced by the community in question was the presence of foreign companies that exploited local natural resources and were backed by the police and army. Many resorts had also been established there because the area attracts international tourism. Some policies established by the resorts disadvantaged local fishermen; for instance, in certain places the latter had lost access to abundant fishing grounds because they were owned by the resorts. The question raised was, How can people be restricted from seeking subsistence in their very homeland? Hadi convinced the villagers that they had the right to fish any place they wanted, arguing that according to the law nobody can possess the sea.

A feeling of resentment developed among the locals toward the companies, mainly because the latter's presence had brought them more difficulties than benefits. As a result of their growing consciousness of their rights, they started to severely criticize the companies, questioning their rights. Tensions rose and conflict between the two sides became a real prospect. At this point, Hadi tried to persuade the people not to use violence to solve their problems. On the other hand, the companies, which were allied with the local authorities, began accusing the NGOs of being troublemakers and of exploiting villagers for their own interests and benefits. Hadi mediated such accusations as a process of learning. He explained the reasons and motives for the companies to propagate such rumors, as in to weaken their opponents and spread disunity between the local population and the NGOs. Hadi was left with a very deep impression of his nine-month praxis with the Bajo tribe of East Kalimantan. He said that it was creativity that made it possible for him to survive and engage in this new environment.

Another experience is recounted by Ina Irawati (2001). She did her nine-month fieldwork in Ranupani, a village located in the area of the Tengger highlands. The people of this village live in the area around the National Park of Bromo Tengger Semeru. Through an interactive dialogue with the local population, Irawati found three major problems in this area: corporate farming that interfered with small-scale growers, unequal rights for local women in the social sphere, and the negative impacts of National Park projects. To deal with these problems, Irawati got involved in the daily activities of women to get a sense of what they felt and thought about their lives. She also initiated a project of redeveloping local fertilizers that had virtually disappeared, in order to end the peasants' reliance on manufactured fertilizers. Further, together with local people, she tried to change, in part, the National Park project so that it benefited the local population.

Impacts on People's Organizations

Social movements and people's organizations are not easy to differentiate because "organizations increasingly resemble episodic movements rather than ongoing bounded actors." The main contribution of Involvement in relation to social movements in Indonesia so far has been to strengthen the local popular organizations. The presence of its alumni in local organizations has opened up new perspectives, particularly with regard to community organizing and Freirean popular education practices.

According to Cahyono (personal communication, 2003), several Involvement graduates have become influential members of many peasant unions, such as the North Sumatra Peasant Union, South Sumatra Peasant Union, Pasundan Peasant Union, Lampung Peasant Union, Sumbawa Peasant Forum, Independent Shoot Peasant Foundation, and

Bogor. They have also become active in many community-based organizations, such as Baelio Ambon, Maur Ohio Wut Development Foundation, Nen Mas Il Foundation, Bina Citra Nusa Foundation, and Hakiki Foundation. In Maluku, almost all the alumni of Involvement have joined the core team of facilitators of Baileo Maluku, the biggest indigenous people's network in Maluku Tenggara.

Some alumni have become involved in environmental movements such as the Keliling Environmental Awareness Group, Delanggu, Klaten; KALIPATRA, Riau, Sumatra; and the Nanimi Wabilisu Foundation, Sorong Papua. Others are now active in advocacy, such as with SAHE Institute for Liberation, East Timor; JKPM Pesantren-Community Network, Wonosobo, central Java; BAR People's Advocacy Agent, Bandung; JPMS Civil Society Empowering Network, Jogjakarta; and FSBS Solidarity Forum of Surakarta Worker. Looking at the roles taken by alumni in local organizations, it is obvious that activist schools like Involvement have made a significant contribution to supporting social movements in Indonesia, not only by strengthening the intellectual capacities of activists that are necessary for responding to the global influence of neoliberalism, but also by strengthening various popular organizations.

Furthermore, Involvement has inspired many NGOs in Indonesia to hold similar kinds of programs. In observing this phenomenon, Cahyono (2002, quoted in Sangkoyo, 2003, p. 4) says: "It should come as no surprise that today there are so many NGOs carrying out Involvement-like programs. It proves that INSIST's ambition in becoming a trend-setter for Indonesian non-government community works out as expected."

Critical Reflection

The method of schooling in Involvement is not without problems, however, as is acknowledged by the inner circle of INSIST. Some argue that there is too much emphasis on abstruse theories at the expense of practical knowledge and skills. According to Amir Sutoko (2002, quoted in Sangkoyo, 2003), a peasant movement activist and senior member of INSIST, the graduates of Involvement are more ideology oriented (read: committed) than non-Involvement alumni. This otherwise positive side is not without risks, though, because "it makes them so busy with themselves and thus weakens considerably their integrative process with the community" (p. 26). What grassroots activists actually need is data-based reflection, or experiment-based knowledge. Even Sutoko argues that the insistence on theoretical knowledge is powerful enough to erase the participants' social memory, a view that nevertheless seems exaggerated.

Toto Rahardjo (2002, quoted in Sangkoyo, 2003) shares the same concern, arguing that Involvement classes do not provide crucial knowledge and skills, like, for instance, how to be a schoolteacher. In his words, "We are still discussing ingredients, not what the final dish that those ingredients will add up to" (p. 25). Locating theoretical knowledge as the main ingredient should not, however, be seen as strategic planning because the graduates will later participate in popular organizations where practical knowledge is definitely needed. This criticism is also based on the assumption that participants are coming from local organizations.

On the other hand, Noer Fauzi (2000) argues that the reason the Involvement course places more emphasis on theoretical instead of practical knowledge is that the curricular framework is designed to radicalize the middle class, or the petit bourgeois type of activist, rather than the local grassroots leader. This is because most participants are former student activists, the educated middle class, or, to use Mansour Fakih's words, "urban activists in T-shirts" (2002, quoted in Sangkoyo, 2003, p. 30). The debate over what elements should be included in the process of learning reflects the plural views of Involvement, particularly among core members of INSIST and program organizers. The philosophical foundation of the program thus remains unsettled. Given this fact, it is important to return to the initial and basic question: What type of alumni does Involvement seek to produce? The answer to this question is significant in determining the content of learning that will be given.

The initial intention behind Involvement was to reschool NGO activists in order to equip them with the intellectual capacities necessary to face the global challenge of neoliberalism. Participants are therefore recruited based on their experiences in social movements. According to the committee organizers, the participants are from the educated middle class or former student activists. For this reason, many of INSIST's key members argue that they do not intend to encourage the graduates of the course to become leaders of people's organizations, such as peasants' or workers' unions, unless they are a genuine part of them. Nor do they intend their graduates to establish their own organizations in order to get foreign funding. The main intention is, by contrast, to strengthen and support local initiatives or people's organizations.

If this is the case then there does not seem to be a crucial issue at hand regarding the emphasis on theoretical knowledge in Involvement. However, the facts are not as they seem. Not all of the participants are former student activists, or the petit bourgeois type of activists. Muhaimin, a participant in the fourth group, for example, did not complete his junior secondary school. He has plenty of practical experience but lacks theoretical knowledge. Similarly, eight of the thirty-one participants of the 2000 class had only graduated from senior high school. In the 2003 class and that which I attended, some had only completed elementary school, such as Samsuddin and Rahmat.

This means that the educational background of the participants varies considerably. Their different social backgrounds have led to a dynamic process of learning. On the one hand, there are participants who ware trained to think rigorously and carefully. When asked to comment on certain social issues, they begin from the very basic questions: What is the problem? Why is it happening? What approach should be taken? Whose theoretical framework can be employed? This is typical of participants who are used to analytical reasoning and have experience in critical thinking.

On the other hand, there are participants who are impatient with this kind of analytical process and who are used to simple forms of analysis. This group usually chooses spontaneous action over abstract theorizing and is usually represented by former street-based activists who participate in demonstrations and in particular those who were involved in opposing and criticizing the Suharto regime in 1998. While the goal of both groups is essentially the same, their methodological approaches are different: one group attempts to analyze problems patiently through theory and analysis, while the other prefers pragmatic and concrete solutions.

The different social backgrounds of the participants should be given special attention because, without assigning too much attention to the issue of variety of educational background, it could nonetheless be helpful to situate some participants in the position of objects. This does not necessarily mean dividing the participants into two groups, but rather ensures that each participant has the opportunity to make his or her voice heard rather than being consigned to the margin. There should be no domination in the process of learning. Indeed, if we look at the reports of class activities, we can see that when discussing certain subjects, particularly in the domain of theoretical knowledge, only a few people are actively involved in the production of knowledge while most others remain silent. The silent majority can be interpreted in two ways: they may be following the discussion and understanding it, or they may really not understand the subject. The second is probably closer to the truth, as can be seen in Anu Lounela's comment (2002, quoted in Sangkoyo, 2003):

> The classroom stage should function as a reflection. This is where the problem of language may interfere. It would almost be impossible for a peasant activist from Wonosobo [a strong foothold of the peasant movement in central Java] to chew Foucault in a short two months. I have been attending Involvement several times. I could see that their gaze was empty. (p. 31)

At this point, it is the task of the facilitator to create a space of learning based on hospitality in order to make participants feel comfortable, free, and self-confident enough to articulate their thoughts, regardless of their different intellectual capacities. The pedagogical process should be run in a participatory spirit in which participants learn from each other, assuming that knowledge and wisdom can be found in books/academia as well as in daily lived experiences. When a hospitable space of learning is created, the learning process will not end up increasing the self-confidence and intellectual capacities of some participants and decreasing those of others.

Another issue that needs to be clarified is whether core members of INSIST should try to produce graduates who go on to lead popular movement organizations. The fact is that many graduates have participated in people's movements. Some of them become key members of these organizations. If these graduates are genuinely part of those communities, there should be no problem. A problem arises, however, when they do not belong to those communities and yet become the leaders. It is important to insist again that it is ethically unacceptable to lead people's organizations if one is not part of that community. The task of the community organizer is not to be the leader of that community, but to nourish, develop, and strengthen its self-reliance. This is, I believe, a basic principle that should be stressed; otherwise, organizing activities will result in increased dependency on the part of the community. Another crucial point, which must be taken into account, concerns the building of curiosity and creativity. Applying theories to everyday practice is not an easy task: it requires considerable creativity. Creativity has helped Hadi to evolve in a completely new environment in the Bajo tribe in East Kalimantan. To survive in a new environment, creativity is a definite requirement, without which it is impossible to ground abstruse theories in concrete reality and make them intelligible to the local people.

However, it is important to note that the task of the activist is not only to ground abstruse theories in concrete reality but also to examine the applicability of theories in practice and to produce praxis-based knowledge, as in knowledge arising from praxis (Conway, 2006). Thus, contrasting theory and local practices allows the activist to know whether the theory is applicable or not, and also to revise and produce it. In the program evaluation report (2002), however, I never found a graduate who spoke about this issue in his/her nine-month fieldwork. Hence it can be argued that the Freirean cycle of reflection-action-reflection has not yet been completed in this instance.

Relating theory and local practice is a crucial issue. It is almost always the case that the point of departure for NGO activists is the theory they find in books. Likewise, the theoretical knowledge transmitted in the Involvement classroom is mostly derived from Western sources. Although these help participants to understand what is going on at the global and national levels, they do not always provide the appropriate lens to perceive a situation at the local level. The major challenge is how to formulate a theoretical framework based on the practices of a local community, or how to make local practices the main reference in formulating a theory. Involvement program, I think, should pay more attention to this indigenization of knowledge. The issue is more than just how to accommodate local content in the pedagogical process of Involvement, but how to theorize indigenous experiences and make them sources of knowledge.

Notes

1. INSIST was established in 1997 as a forum of communication amongst NGO activists who have experience in conducting social transformation at grassroots levels through community organization and popular education. It was initiated to function as a medium for conceptualizing and systemizing practical experiences that could serve as guidance for young social activists involved in social movements and transformation. Since its establishment, INSIST has made significant contributions not only in terms of developing critical discourses and alternative perspectives but also in terms of supporting people's organizations and social movements in Indonesia.
2. KEPA, (*Kehitysyhteistyön Palvelukeskus*), or the Service Centre for Development Cooperation, is a network of more than 200 Finnish NGOs interested in development network and global issues. Its mission is to transform the structures that cause and sustain inequality in the world. In addition to its work in Finland, KEPA has field offices in Mozambique, Nicaragua, and Zambia. KEPA has been involved in a unique partnership with INSIST since it was founded, with one KEPA staff member working permanently at the INSIST office in Jogjakarta.

References

Conway, J.M. (2006). *Praxis and politics: Knowledge production in social movements*. New York: Routledge.

Djamhari, C. (1996). *Privatization of state controlled enterprises in Indonesia (1983–1996): Policy and practice.* Master's thesis, Department of Sociology, McGill University, Montreal, Canada.
Eldridge, P.J. (1995). *Non-government organizations and democratic participation in Indonesia.* New York: Oxford University Press.
Fakih, M. (1996). *Masyarakat sipil untuk Transformasi Sosial: Pergolakan Ideologi LSM Indonesia.* Jogjakarta: Pustaka Pelajar.
Fauzi, N. (2000). *Laporan Kegiatan Kelas.* Jogjakarta: INSIST.
Freire, P. (1971). *Pedagogy of the oppressed.* New York: Herder and Herder.
Freire, P. (1985). *The politics of education: Culture, power, and liberation.* South Hadley, MA: Bergin & Garvey.
Hadad, N. (2003). *Water resource policy in Indonesia: Open doors for privatization.* Retrieved November 2006 from http://www.jubileesouth.org/news/EpZZZZFyyAFYXArDht.shtml
Irawati, I. (2001). *Laporan Kegiatan Kelas.* Jogjakarta: INSIST.
The Jakarta Post. (1997, October 31).
Laporan Kegiatan Kelas. (1999). Jogjakarta: INSIST.
Laporan Kegiatan Kelas. (2000). Jogjakarta: INSIST.
Sangkoyo, H. (2003). *In good company: INSIST-KEPA collaboration, 1992–2002.* Unpublished report.
Tamher, D. (2001). *Laporan Kegiatan Kelas.* Jogjakarta: INSIST.
Topatimasang, R. (2000). *Laporan Kegiatan Kelas.* Jogjakarta: INSIST.

CHAPTER 8

Nonformal Education, Economic Growth and Development: Challenges for Rural Buddhists in Bangladesh

Bijoy P. Barua

Introduction

Nonformal education is considered to be an important precondition for social change and for alleviating poverty in rural societies in developing countries. In other words, nonformal education is a necessary and indispensable component of the rural development programs of recent times. It is an organized educational activity focused on individual technical skills and directed toward social and economic change for rural development (Fink, 1992). It promotes a *trickle-down* approach instead of a *bottom-up* approach. Over the years, nonformal education has tended to adopt the agenda of cultural homogenization in the name of development. This education has deep colonial roots that intend to create materialistic values and greed. Villagers have been continually educated through media networks such as radio, television, and newspapers and through rural extension workers on issues of agricultural development (Kashem, Halim, & Rahman, 1992), individual freedom, democracy, industrialization, urbanization, growth of the Western world, and economic growth. Education in this form is regarded as a method of depositing information in the villages. In other words, nonformal/extension education disengages the rural people from their native land and from their own knowledge base in the name of growth and development. Over the years, the process and policies of education for agricultural development have been modified and re-organized through rules and regulations in order to control the village communities. In contrast, Buddhist experiential learning engages with integrated farming for biodiversity and a self-reliant community through a nonviolence approach (Barua, 2004). Under this perspective, nonformal education appears to contradict the popular Buddhist way of learning. Buddhist experiential learning is ecocentric and is not preoccupied with the private fortune of the individual; rather, it gives awareness of all living beings and creatures. It is more concerned with a reflective learning practice, democratic pluralism, an ecocentric environment and society. Its knowledge is generated through a process of critical awareness. In this learning process, *engagement* and *enlightenment* are considered to be the key elements for social transformation and for promoting the well-being of all living creatures. In this effort, the highest priority is given to the notion of *bahujanahitaya, bahujanasukhaya* (for the happiness and welfare of all). It rejects the notion of authoritarianism, consumerism, and large market economy. It does not believe in imposition (Barua, 2003, 2004). However, this learning has been challenged due to the imposition of a *neoliberal* agenda.

This chapter will look closely at the nonformal/extension education for community and social development in a Buddhist village within the present socioeconomic context of Bangladesh. I will argue that nonformal/extension education promotes only monoculture economic growth and ignores cultural pluralism. I will also argue that Buddhist education allows the liberation of the mind through the extinction of *lobha* (greed), *dosa* (hatred), and *moha* (delusion). This leads to the building of sustainable communities.

The data for this chapter was contextualized through applications of Buddhist experiential learning, knowledge, and practice in Asian Buddhist societies in general and in Bangladeshi societies in particular. This chapter is based on qualitative research done in the village of Muni, situated in the northeast of the Chittagong District of Bangladesh, during fall 2001 and winter 2004. The village lies at the foot hills and is surrounded by forests. It has 411 households and the economy is based mainly on agriculture, which primarily involves rice cultivation in the valleys and gardening in the forests. This self-sufficient village economy is deeply rooted in the lands, ponds, and forests. Drastic economic changes have taken place since the building of the Kaptai Dam and Hydro Power Station in the Chittagong Hill Tracts District on the

Karnaphuli River in 1960, which is 15 kilometers upstream (east) of the village of Muni (Barua, 2004).

In the following sections, I will look at the context of Bangladesh and Buddhists and the historical context of extension education in the country, and then review the agricultural development process in Bangladesh. I will then critically examine the agricultural development in a Buddhist village through narratives of community farmers/members. Finally, I will draw a conclusion from the perspective of Buddhist experiential learning and development. For convenience of discussion, the terms *nonformal* and *extension* education will be used interchangeably.

The Context of Bangladesh and Buddhists

Bangladesh is an independent nation-state with a population of 134.8 million, of which 69.1 million are male and 65.7 million are female (Bangladesh Bureau of Statistics, 2007). It is the third largest Muslim-inhabited country in the world. Muslims represent 88.3 percent of the country's total population, and Buddhist people constitute about 0.6 percent of the total population (Barua, 2004). About 75 percent of the population is dependent on agriculture (Bangladesh Bureau of Statistics, 1999). The introduction of a trade liberalization policy brought Bangladesh into the network of the global market economy (Barua, 2004). This development paved the way for the establishment of export processing zones in large cities, to where the majority of workers have migrated from the rural areas.

There are four distinct Theravada Buddhist communities in Bangladesh: Baruas, Chakmas, Rakhaines, and Marmas. They live in the southeastern part of the country. This concentration is mainly due to their geographical closeness to the neighboring Buddhist country of Myanmar (Burma). Among these Buddhist communities, the Baruas and Rakhaines live mainly in the Chittagong, Cox's Bazar and Patuakhali districts. There are about 635 monasteries in Bangladesh (Barua, 2004). For the Buddhists, the monasteries are centers of learning, spiritual training, and social development in the villages where the members of the community gather, meditate, celebrate festivals, and discuss sociocultural issues through dialogue and story (Barua, 2004). The monasteries are not only "confined to the philosophical and psychological aspects of the religion but extend to the field of social service and the cultivation of self-discipline" (Dhammananda, 1996, p. 94). During the last ten years, the socioeconomic conditions in the village have changed drastically. The collapse of the rural economy in the Buddhist villages has dislocated people from their own land, minds, and spirituality. At the same time, the situation of Buddhists in Bangladesh is conditioned by their minority status and therefore their participation in the development process has been neglected.

Historical Context of Nonformal/Extension Education in Bangladesh

Nonformal/extension education is perceived as a provider of modern values in an effort to change the cultural values of the rural people of Bangladesh. This education tends to focus on technical skills, high growth, and the market (Kashem et al., 1992) rather than on reflective and contemplative learning practices. Nonformal/extension education and agricultural development have worked hand in hand to change economic conditions and introduce materialistic values in rural communities. This model of agricultural development was institutionalized in 1870 to regulate the rural economy in Bangladesh. To serve this goal, a separate Department of Agriculture was created in 1906. In time, extension education for agricultural development was initiated to disseminate information to the rural farmers. The education mainly focused on a process of centralization of knowledge in the cities rather than a decentralization of knowledge within the cultural context. Extension education was market driven and profit oriented. It also worked hand in hand with consumerism. For the past five decades, Bangladesh has experimented with various approaches to nonformal education in an attempt to change the life of the rural people without necessarily embracing location-specific knowledge and ethical issues. An organized nonformal education program began in 1956 through the Village AID (Village Agriculture and Industrial Development) programs to educate and mobilize rural people and improve their socioeconomic conditions. However, after the establishment of the Bangladesh Academy for Rural Development (BARD) in 1959, an experimental nonformal education program was initiated in the Comilla District in order to mobilize rural farmers. In the same way, Bangladesh Agricultural Development Corporation (BADC) was founded in 1961 to bring about major changes in agricultural development with the goal of promoting modern technology, deep tube-wells, tractors, and fertilizers. Similarly, the government established an Adult Education Division in 1963 under the Directorate of Education to enhance the efforts of rural development and extension education. These programs mainly focused on increase in agricultural production and economic growth.

In 1977, an extension and research project was launched with the support of the World Bank in Bangladesh. This initiative was promoted to industrialize the farming system through a cadre of technocrats. This large-scale extension education process and research service focused on a training and visitation system to increase production through plantation of monocrops and the distribution of fertilizers and pesticides (Ali, 1973; Halim, 1995; Momin, 1984; Qadir & Balaghatullah, 1985). This effort has been advocated by development agencies in Bangladesh due to the success of the green revolution in the North. Education for agriculture modernization tends to use the transfer technology method rather than the action-and-reflection process. The transfer technology method ignores the diverse knowledge systems and experience of the rural farmers in Bangladesh. This method has become a dominant tool in regulating the farmers' way of life in the name of modernization. It emphasizes a capital- and energy-intensive approach. It promotes high technology in a labor-intensive society and undermines appropriate technology and culture (Barua, 1999). The growth-oriented economy is confined to matters of land control, technology, production, cash, finance, and profit. This method does not look at any of the connections between life and nature. Land is considered as a commodity with a monetary value. The spiritual connection of the farmers to their land is ignored. The farmers' knowledge is considered as backward and illogical. Rural farmers are portrayed as inferior in an age of transfer technology within the framework of agricultural development and the growth model (Chamber, 1993). These notions are based on Western economic theory and culture rather than on the local economic logic and realities (Barua, 2004). Consequently, the rural economy has been replaced by a market-driven economy. Centralized urban control has been imposed on the rural farmers for the purpose of increased agricultural production. The decentralized decision-making power of farmers has been eroded. Over the years, modern agriculture has essentially displaced the indigenous varieties of crops. For example, Bangladesh has lost 7,000 local varieties of rice (Environmental Coalition of NGOs and Association of Development Agencies in Bangladesh, 1992). Such modernization has not only marginalized the rural farmers and created dependency on the external inputs of the so-called expert, but it has also shown disrespect to the land, nature, and local ethical values. It has also dehumanized people and denigrated local culture. Over the years, extension education was designed to desensitize the rural communities in the name of growth and agricultural development, which practically created dependency on the inputs of transnational companies. In other words, the extension education model has worked toward the colonization of the rural communities to ensure more profit for transnational companies.

Nonformal education for agricultural modernization also advocated dams for irrigation facilities, and chemical fertilizers and pesticides for economic growth and progress with the assistance of multilateral and bilateral agencies. This has been the trend in the last five decades with the aim of bringing about high-technology facilities for higher productivity and growth. The dams, chemical fertilizers, and pesticides were promoted for industrialization of the rural economy of Bangladesh. This Western economic model has not improved the quality of life but has created an environment of social inequality and crises in the villages (Barua, 2004). "The role of agriculture has declined over the last twenty to twenty-five years" (Huq, 2001). In fact, net cultivable land declined from nearly 8.2 million hectares in 1983–1984 to about 7.2 million hectares in 1995–1996. Employment opportunities in the agricultural sector have declined from 86 percent in 1974 to 73 percent in 1996. Rural people are more inclined toward the city for their economic survival. The percentage of people living in villages has declined from 90 percent in 1980 to 80 percent in 1995 (Huq, 2001).

In the following section, I will critically examine extension education and agricultural development and construct knowledge from a Buddhist perspective. Specifically, I will look at how a Buddhist village has been mobilized and what challenges they encounter in the socioeconomic and political context of Bangladesh within the framework of international development.

Economic Growth and Development: Displacement of Buddhist Economy and Ethics

The global economic agenda and cultural penetration from the West disregard Buddhist economic values and traditions. The agricultural modernization scheme and economic growth model are not designed within the context of the people, culture, and environment of the village considered in this case study. This trend neglects the Buddhist practices of simplicity and nonviolence. The Buddhist notion of economic development avoids massive expansion, high growth, and equipment, which tend to control human beings rather than serve them (De-Silva, 1998; Schumacher, 1973). A Buddhist economy is more concerned about the generation and utilization of resources for the well-being of all. In other words, this type of economy does not encourage the preservation of unlimited goods for one's benefit (Schumacher, 1973 Sivaraksa, 1990;). Buddhist economy emphasizes the minimum consumption of natural resources to maintain harmony with nature and society, which also helps to sustain a locally self-sufficient economy (Schumacher). Poverty is not accepted in an ideal Buddhist society. The foundation of a Buddhist economic culture is based not on the accumulation of material wealth but in the refinement of human quality. The production of local resources for local needs is the most appropriate mode of economic life in a Buddhist society. The Buddhist notion of agriculture and economy is based on five pillars or principles, which include spirit or *dhamma,* natural balance, appropriate technology, community life (monastery/culture), and economic self-reliance. If these five interconnected principles are preserved within production patterns, the society tends to be sustained without external dependency (Wasi, 1988). However, the model of a growth economy unfortunately neglects the realities of Buddhists in the village in order to nurture and promote the agenda of multinational or transnational companies. As a result, Buddhist-integrated agriculture has collapsed. Over the last five decades, the displacement of the rural economy has practically eradicated employment opportunities in the village (Barua, 2004). Rona, a village farmer, expressed it this way:

> Our village economy is fragile now. We have lost our economic self-sufficiency. Our economic condition is in bad shape, we do not now see the economic environment in the village we had in the past. For this reason, our villagers are compelled to leave the village for their survival. (cited in Barua, 2004, pp. 168–169)

My own survey estimated that 343 people have temporarily migrated from the village of Muni to the city of Chittagong in order to work in export processing zones. Most of these migrants are typically adult males and females. Although this movement toward the city was to escape from the economic hardships of the village, these people could not avoid the terrible working conditions of the city and its industries. Hundreds of adult members of the village have also migrated to Japan, South Korea, Europe, and North America. This migration is obviously linked with economic insecurity in the case-study village (Barua, 2004). In recent times, the villagers have often organized their wedding ceremonies in the city due to the shift in economic context and changing cultural environment of the village. Such dislocation of people is disastrous for the community and the society as a whole. The majority of community members have turned to the cities, "having left the land and their local economy to end up in the shadow of an urban dream that can never be realized" (Norberg-Hodge, 1991, p. 149). Obviously, this has also led to moral degradation and the people's displacement from their own roots and happiness. "Jobs in remote workplaces are not truly 'economic'; for they erode the villagers' true security, which is inseparable from their family and community relations" (Macy, 1994, p. 160).

The Buddhist practice of economic development is much different from a growth-oriented economy. The Buddhist economy aims to develop awareness among people to control materialistic desire and greed. It is meant to harmonize people with the natural environment and their social conditions.[1] Humans are an integral part of the natural system in Buddhism. Self-sufficiency is regarded as indispensable to the sustainable development of society in a Buddhist economy. It does not seek to create delusive *pseudo-desires* amongst the people. Dependence on imports from far away and exports to unknown places and distant peoples is extremely uneconomical from the Buddhist point of view. As community members are heavily engaged in the production of new high-yielding plant varieties and packages of inputs, their lives are controlled by external forces (Barua, 2004). A Buddhist farmer Mona describes this:

> Nowadays, it is impossible to cultivate such modern variety without fertilizers and pesticides. If one used them, he would have to distribute them every time. For this, we have to spend money. It is costly and expensive. Every year, we have to increase the amount of fertilizers and pesticides. (cited in Barua, 2004, pp. 162–163)

Similarly, Chan, a village farmer, states:

> I work with a power tiller all year without the rest and recreation necessary for survival. However, all of my ancestors' agricultural lands and forests have gone out of my hands at the cost of HYV, fertilizers, pesticides and mechanization. (cited in Barua, 2004, p. 76)

This trend toward a growth economy tends to help the city-based dealers, traders, and multinational companies rather than the rural farmers. A similar situation was observed by the scholar Kortén (1998) while reflecting on the conditions of agricultural modernization.

> Through costly experience we are learning that the term free market is a code word for giving capitalism a free hand to colonize the living resources

> of the planet for short-term financial gain at the expense of human freedom, prosperity, and even of the market itself. (p. 138)

The transnational companies are mostly occupied with the package of technologies, such as seeds, fertilizers, and pesticides. For example, Monsanto is actively engaged in selling its products to the rural farmers through its partner in Bangladesh (Huq, 2001). These inputs are too expensive for the farmers. In many instances, farmers end up in debt and eventually lose their land. In the past, the farmers of the case-study village were seldom in debt and were very independent. They used their own seeds for cultivation and cow dung as fertilizer (Barua 2004).

Over the years, Buddhist ethics have been neglected in the village due to the expansion of a free market economy. "In a free market open economy, the religious and spiritual heritage of our societies have been brushed away leaving room for the competitive and possessive instincts of individuals to flourish" (Ariyaratne, 1996, p. 236). The open market increased the overall use of inorganic fertilizer and pesticide inputs on the farmlands, as is required for modern intensive agricultural production (Barua, 1999). The national data of Bangladesh reveal that the use of chemical fertilizers has increased from 2 million metric tons in 1990–1991 to 3.02 million metric tons in 1995–1996. The price of fertilizers has increased four times in the last twenty years. Accordingly, the cost of pesticides has also increased ten times in this same period. Despite the great increase in the use of fertilizers and pesticides, the production of *boro* (winter rice) in Bangladesh actually declined 22.2 times in the last ten years (Association of Development Agencies in Bangladesh, 1997). Further, the production of HYV rice per acre in 1986 was reduced by 10 percent from the 1972 level, despite a 300 percent increase per acre in the use of chemical fertilizers and pesticides (Environmental Coalition of NGOs and Association of Development Agencies in Bangladesh, 1992). Evidently, inorganic fertilizers have displaced the locally available organic fertilizers in the village. Such a shift has not only created dependency on external inputs in the rural economy but has also increased the destruction of nature and environment. More distinctively, the use of chemical fertilizers and pesticides has been involved in the killing of insects and other creatures. Chan mentioned:

> The pesticides we use in our farmlands today obviously kill insects and animals. The use of pesticides definitely contradicts Buddhist ethics. In fact, the Buddha never encouraged people to kill any living creatures. (cited in Barua, 2004, p. 164)

In the name of growth, the agricultural development trend ignores the Buddhist vision of nonviolent acts and interconnectedness. Over the years, farmers have been involved in causing harm to nature and all other insects and small creatures through the instigation of multinational companies. Such practice and behavior is not acceptable in Buddhist societies. Hannah, a Vietnamese Buddhist monk and scholar, stated critically:

> The excessive use of pesticides which kills all kinds of insects and upsets the ecological balance is an example of our lack of wisdom in trying to control nature. Economic growth, which devastates nature by polluting and through the exhaustion of non-renewable resources, renders the earth impossible for beings to live on. Such economic growth may temporarily benefit some humans but in reality disrupts and destroys the whole of nature to which we belong. (Wasi, 1988, p. 43)

Using chemicals to enhance growth for economic benefit is indisputably a violent act (Hanh, 1996; Wasi, 1988). Buddha was always concerned about all living creatures. For example, he made a strict rule against traveling through the rural earthen roads and agricultural land during the rainy season in order to avoid possible injury to worms and insects. The concept of *vassavasa* (rain retreat for three months) was thus initiated by Buddha in order to preserve the first precept of noninjury and to cultivate sympathetic love for all creatures (Barua, 2004).

The modern technology-based irrigation projects have also ignored the local autonomy, appropriate technology, and unity of communal irrigation systems in the village. For example, the building of the Kaptai Dam in the Chittagong Hill Tracts in 1960 created disasters in the natural irrigation facilities and for the ethnic minorities. In the case of the village Muni, the Kaptai Dam not only blocked the flow of the river water and natural irrigation facilities but also replaced the local irrigation technology. This replacement created a dependency on urban-made technology and city-based experts. Overuse of the water resource has also created ecological disasters such as increased salinity on the farmland. For these reasons, the agricultural lands have often become noncultivable (Barua, 1999). Modern irrigation systems have also contributed to a major health crisis, which is reflected in increased cases of schistosomiasis and malaria among the people. These diseases have increased due to salinization, alkalinization, and water logging in farmland areas (Economic and Social Commission for Asia and the Pacific [ESCAP], 1985). In Bangladesh, "more than three million hectares of lands are affected by salinity" (Islam

& Sadeque, 1992, p. 75). Mechanization and modernization processes are found to be inappropriate for the diverse systems of agriculture. However, "wisdom demands a new orientation of science and technology toward the organic, the gentle, the non-violent, the elegant and beautiful" (Schumacher, 1973, p. 27). The denial of wisdom in the name of progress and scientific development is an encouragement of devastation and crisis in the natural environment. Buddhist texts not only reflected about rice cultivation but also referred to appropriate irrigation techniques in the local context (Barua, 2004). The theory of development must be of practical relevance to the context, culture, and environment of villages for the benefit of the rural communities. For example, the construction of a 1,000 tanks and more than 600 miles of irrigational channels in ancient Sri Lanka clearly proved that the technology built on indigenous knowledge was appropriate to the local environment and culture (Gamage, 2001). Despite this, development planners and experts have been engaged in more destruction in the name of scientific technology and development for economic growth.

Monoculture: Displacement of Diversity

The promotion of monocrop has become a dominant model within the framework of agricultural development in Bangladesh. This model is based on a Western economic theory (Norberg-Hodge, 1991). Over the years, the growth model has failed to address the issues of biodiversity and cultural knowledge. The Buddhist villages are no exception. For the Buddhists, agriculture is not only a basic need, but it also gives them the opportunity to live in harmony with the natural environment. It allows farmers to independently sustain their lives in their natural setting. The origins of knowledge and life are inseparable in Buddhist education. If knowledge is devised without a knowing of nature, it can be harmful to society and environment (Barua, 2004). In Buddhist experiential learning, justice is not limited to human beings but proceeds beyond human rights to embrace the rights of all living creatures, nature, and the environment for sustainable development. Buddhist learning is "ecocentric rather than theocentric" (Sponsel & Natadecha-Sponsel, 1993, p. 78). Its notion of development is profoundly rooted in the concept of *ahimsa* (nonviolence). In considering the well-being of the society, "the Buddha did not take life out of the context of its social, political and economic background; he looked at it as a whole, in all its social, economic and political aspects" (Rahula, 1994, p. 24). Buddha's education was centered on *mindfulness* and *right action*. The practice of mindfulness helps us to look deeply into our problems in order to initiate the right action for daily life. The practice of mindfulness is not an escape from the social environment; rather, it prepares us for reentry into the community with deep commitment to ensure better quality of service to all living beings (Hanh, 1996). Therefore, Buddhist education promotes cultural diversity and ecological balance. However, extension education is designed to cultivate monoculture or cash crops for individual monetary benefits as opposed to the long-term benefits for the local culture as a whole. In truth, such a notion of economic development is inappropriate for the Buddhist communities. Today, diversity of rice no longer exists in the case-study village (Barua, 2004). Plant life has turned into a nonrenewable resource (Shiva, 2000). In the past, 154 varieties of rice were available in the greater Chittagong area. These used to be planted by the farmers on a regular basis without any fertilizers or pesticides. Crops were produced in different seasons and at different times. The indigenous varieties ensured food for the people and also allowed the rural people to celebrate their diverse socioreligious and cultural festivals (Barua, 2004). Now, local varieties such as *gacca* (deep-water aman) have disappeared from the farmland of Buddhist villages. This variety of rice was developed by farmers for centuries in the context of a flood-prone deltic environment. The deep-water variety of rice could grow up to 18 feet tall in floodwaters (Barua, 2004).

The green revolution technology has also reduced production of winter crops in Bangladesh. As a result, pulses, mustard seed and oil, and other varieties have been displaced from the farmland of Buddhist villages because of the increased emphasis on high-yielding varieties of rice (Barua, 2004). The above varieties of crops are a main source of protein for the people in villages. Moreover, mustard oil can be used in cooking and also as an effective herbal medicine in the indigenous health care system. It is specially used for healing muscular and joint ailments. At the same time, it is used to light lamps, in the celebration of festivals, and also as an environmental purifier to prevent disease and to control pests (Shiva, 2000). Mustard, along with all these other varieties of crops, virtually disappeared from the soil of the village due to the strong trend toward monocrop production. The failure of the green revolution has resulted in a turn toward biotechnology (Laird, 2000). This shift in technology itself proves that agricultural modernization is not sustainable in this context and environment. Debates about biotechnology have already become intense.

Concluding Remarks

In this chapter, I have critically examined nonformal/extension education in the context of Buddhist society and culture. Over the years, agricultural development schemes have ignored the Buddhist vision of nonviolent acts and interconnectedness in the name of growth. The case study demonstrates that extension education policies have pursued the path of the colonial that endorses centralized control and serves the interests of urban traders and transnational companies. Centralized control has failed to address biodiversity, ecocentric development approaches, and ethical (moral) issues while implementing extension education and agricultural development. The centralized policy has also undermined cultural knowledge for the sake of material growth and gain. Further, the case study clearly indicates that the extension education process and agricultural modernization have not been able to apply a Buddhist reflective approach to understand local knowledge, wisdom, and economy. The Buddhist learning pedagogy does not believe in any imposition. This learning process is creative and reflective and attempts to address the context and environment. More importantly, Buddhist learning sensitizes the people to decolonize the mind for the well-being of all creatures and the environment. "Knowledge gained through spiritual means can serve economic as well as psychological needs" (Huq, 2001, p. 24).

The case study also clearly confirms that knowledge without wisdom is harmful to society and the environment. It is also obvious that the theory in education and development must be relevant to the realities and context of the people. Local knowledge, economy, culture, and wisdom must be promoted for sustainable development. Given this fact, we must find a middle path based on the notion of a Buddhist education in order to protect the earth and promote biodiversity. For Buddhists, it is important to preserve the first precept of noninjury and to cultivate sympathetic love for all creatures. Buddhist cultural knowledge offers us a way to think in alternative terms to articulate the daily experiences of a community in order to prepare better citizens for a society. It promotes a simple life through a spiritual balance to control greed and environmental destruction in society for sustainable resource generation. Its notion of agriculture and economy is based on five pillars or principles, including (1) spirit or *dhamma,* (2) natural balance, (3) appropriate technology, (4) community life (temple/culture), and (5) economic self-reliance. These ethical principles offer integrated agriculture. They also allow the communities to empower themselves for sustainable living without external dependency and imposition in order to attain liberation. These principles reject the notion of authoritarianism, consumerism, and large market economy. More importantly, Buddhist teachings advocate that the good governance of a society demands the responsibility to create favorable conditions and environments for the welfare and happiness of the people through a *sangha* (democratic forum) of community. The issue I raise here is not the imposition of Buddhist approach upon others as panacea. Rather, it is an invitation to critically think about alternative development approaches in the context of Buddhist societies that respect right livelihood, peace, and nonviolence for sustainable development.

Note

1. Based on these learning experiences, several organizations have been involved in implementing nonformal education and rural development activities in Thailand and Sri Lanka.

References

Ali, A. (1973). *Rural development in Bangladesh.* Comilla, Bangladesh: BARD.

Ariyaratne, A.T. (1996). *Buddhism and sarvodaya: Sri Lankan experience.* Delhi: Sri Satguru Publications.

Association of Development Agencies in Bangladesh. (1997). *Adhuna Development Quarterly, 6* (1 & 2).

Bangladesh Bureau of Statistics. (1999). *Statistical yearbook of Bangladesh, 1999.* Dhaka, Bangladesh: Government of the Peoples' Republic of Bangladesh.

Bangladesh Bureau of Statistics. (2007). *Statistical pocketbook of Bangladesh, 2005.* Dhaka, Bangladesh: Government of the Peoples' Republic of Bangladesh.

Barua, B.P. (1999). *Non-Formal education and grassroots development: A case study from rural Bangladesh.* Unpublished master's thesis, Concordia University, Montreal, Canada.

Barua, B.P. (2003). *Buddhist experiential learning and contribution to adult education in an era of globalization.* Proceedings of the Twenty-second Annual National Conference of Canadian Association for the Study of Adult Education, Dalhousie University, Halifax, Nova Scotia, Canada, 1–5.

Barua, B.P. (2004). *Western education and modernization in a Buddhist village of Bangladesh: A case study of the Barua community.* Unpublished doctoral dissertation, Department of Sociology and Equity Studies in Education, University of Toronto, Canada.

Chamber, R. (1993). *Challenging the professions: Frontiers for rural development.* London: Intermediate Technology Publications.

De-Silva, P. (1998). *Environmental philosophy and ethics in Buddhism.* London: Macmillan Press.

Dhammananda, K.S. (1996). *Daily Buddhist devotions.* Taipei, Taiwan: The Buddha Educational Foundation.

Economic and Social Commission for Asia and the Pacific. (1985). *State of the environment in Asia and Pacific,* Vol. 1.

Environmental Coalition of NGOs and Association of Development Agencies in Bangladesh. (1992). *Environment and development: Bangladesh NGOs perspective on policy and action.* A position paper for the United Nations Conference on Environment and Development, Rio de Janeiro, Brazil.

Fink, M. (1992). Women and popular education in Latin America. In N.P. Stromquist (Ed.), *Women and education in Latin America: Knowledge, power and change* (pp. 171–193). Boulder, CO: Lynne Rienner.

Gamage, D.T. (2001). A monastery system of higher education: Twenty-three centuries of Sri Lankan experience. *Asian Profile, 29*(1), 31–41.

Halim, A. (1995). Diffusion of agricultural technology in Asian villages: Bangladesh. In J. I. Bakker (Ed.), *Sustainability and international rural development* (pp. 203–219). Toronto: Canadian Scholar's Press.

Hanh, T.N. (1996). *Being peace.* Berkeley, CA: Parallax Press.

Huq, H. (2001). *People's practices: Exploring contestation, counter-development and rural livelihoods.* Dhaka, Bangladesh: Community Development Library.

Islam, A.M., & Sadeque, Z.S. (1992). *Environment and natural resource management in Bangladesh.* Dhaka, Bangladesh: Sociological Association.

Kashem, M.A., Halim, A., & Rahman, M.Z. (1992). Farmers' use of communication media in adopting agricultural technologies: A farm level study in Bangladesh. *Asia-Pacific Journal of Rural Development, 2*(1), 94–112.

Korten, D.C. (1998). *The post-corporate world: Life after capitalism.* San Francisco: Berret-Koehler and Kumarian Press.

Laird, J. (2000). *Money, politics, globalization, and crisis. The case of Thailand.* Singapore: Graham Brash.

Macy, J. (1994). Sarvodaya means everybody wakes up. In C. Whitemyer (Ed.), *Mindfulness and meaningful work* (pp. 148–164). Berkeley, CA: Parallax Press.

Momin, M.A. (1984). Institutional building and rural development in Bangladesh. *Journal of Local Government, 13*(1), 45–61.

Norberg-Hodge, H. (1991). May hundred plants grow from one seed, the ecological tradition of Ladakh meets the future. In M. Batchelor & K. Brown (Eds.), *Buddhism and ecology* (pp. 41–54). England: Cassell.

Qadir, S.A., & Balaghatullah, M. (1985). The role of education in integrated rural development in Bangladesh. In D. Berstecher (Ed.), *Education and rural development: Issues for planning and research* (pp. 181–225). Paris: UNESCO.

Rahula, W. (1994). Buddhism in the real world. In C. Whitemyer (Ed.), *Mindfulness and meaningful work: Explorations in right livelihood* (pp. 23–27). Berkeley, CA: Parallax Press.

Schumacher, E.F. (1973). *Small is beautiful: Economics as if people mattered.* New York: Harper & Row.

Shiva, V. (2000). *Stolen harvest: The hijacking of the global food supply.* Cambridge, MA: South End Press.

Sivaraksa, S. (1990). True development. In A.H. Badiner (Ed.), *Dharma gai: A harvest of essays in Buddhism and ecology* (pp. 169–177). Berkeley, CA: Parallax Press.

Sivaraksa, S. (1992a). *Seeds of peace, a Buddhist vision for renewing society.* Berkeley, CA: Parallax Press.

Sivaraksa, S. (1992b). Buddhism and contemporary international trade. In K. Raft (Ed.), *Inner peace, world peace: Essays on Buddhism and nonviolence* (pp. 127–137). New York: State University of New York Press.

Sponsel, L., & Natadecha-Sponsel, P. (1993). The potential contribution of Buddhism in developing an environmental ethic for the conservation of biodiversity. In L. Hamilton, (Ed.), *Ethics, religion and biodiversity, relations between conservation and cultural values* (pp. 75–97). Cambridge, UK: The White Horse Press.

Wasi, P. (1988). Buddhist agriculture and the tranquilty of Thai society. In S. Phongphit & R. Bennoun (Eds.), *Turning point of Thai farmers* (pp. 1–43). Bangkok, Thailand: Moo Ban Press.

CHAPTER 9

East Meets West, Dewey Meets Confucius and Mao: A Philosophical Analysis of Adult Education in China

Shibao Guo

Introduction

The Chinese emphasis on education is legendary, and adult education has been part of this legendary history since the beginning of its 5,000 year-old civilization. Among many great thinkers who have provided philosophical visions and guidance in the history of Chinese adult education, three philosophers are most influential: Confucius, John Dewey, and Mao Zedong. Confucius is one of the earliest recorded adult educators. During his life, 3,000 disciples studied with him; 72 of them became very distinguished. Much of his time was spent in teaching adults, particularly during the years he spent traveling across the country. For centuries to come following his death, Confucius's influence remained strong in Chinese history by way of reproduction of his classics in the civil service examinations. John Dewey was an American philosopher who visited and lectured in China from 1919 to 1921. During his stay, he traveled extensively in China and lectured on philosophy, in general, and the philosophy of education, in particular. He was among the first to challenge Confucianism. His impact on recent Chinese adult education was immense, particularly from the 1920s to 1940s, before the People's Republic of China was established. With the birth of the new republic in 1949 came the era of Mao Zedong. Many people in the world know Mao as a communist revolutionary leader. In fact he was also one of the most important social and ideological forces in modern Chinese education, if not the only one. As early as 1917, Mao started the first evening school for poor workers to raise their literacy level in his hometown, Human (Xiao, 2003). Under Mao, the aim of adult education has shifted from producing a class well versed in Confucian classics, to the training of new socialists who serve proletarian politics combined with developing a productive labor force (Guan, 1987; Yao, 1987).

The aim of this chapter is to examine the contributions of these great thinkers to the development of adult education in China. It is also to explore how these various ideological forces coexisted and interacted in bringing China's adult education system to its current shape, and, furthermore, to determine which force(s) exert the strongest influence on China's adult education. This chapter is divided into four parts. Part one starts with a brief introduction to the history of Chinese education, especially of Confucian education. Part two investigates Dewey's visit to China and his influence on education in China. The third part compares Dewey with Mao, who founded the communist China and led the campaign against Dewey's progressivism in the 1950s. The final part examines the various social and ideological forces that coexist and interact with each other and determines which force(s) exert the strongest influence on China's current education system.

Confucius and the History of Chinese Education

This part will briefly review the historical development of Chinese education, especially Confucian education. First, we will examine Confucius and his major philosophy and educational ideology. Next, we will discuss education after Confucius till the last dynasty was overthrown in 1911.

Confucius (551–479 B.C.)

Confucius was the greatest and most highly revered of all traditional Chinese philosophers. He was not only a great philosopher but also a great adult educator. Chinese education since the Han Dynasty (207 B.C.–A.D. 220) has been greatly influenced by him. It is beyond the scope of this chapter to discuss Confucius in great detail; however, it is necessary to introduce some of his most important philosophical and educational ideas to have a better understanding of this sage.

Confucius was born into a small aristocratic but impoverished family in the Lu Kingdom (the present Shandong Province). His father died when he was three years old. He was brought up by his mother, an able and virtuous woman, and the family education provided by her proved to be very important for Confucius throughout his life. When he was fifteen years old, he decided to devote his life to learning. He was very diligent in his studies and adept at learning from people around him. He said: "Among three people walking with me, one must be my teacher" (Pound, 1956, p. 107).

One of Confucius's most important philosophies was "Ren" or benevolence. Its fundamental meaning was "to love others"; it also meant "self-cultivation." Benevolence also included filial piety and loyalty. The purpose of advocating benevolence was to establish a stable and peaceful social order. Confucius's other important ideologies were rites and mean. He said, "Look not at what is contrary to propriety; listen not to what is contrary to propriety, speak not what is contrary to propriety; make no movement which is contrary to propriety" (Pound, 1956, p. 145). Confucius saw mean as the perfect virtue, taking it as the supreme standard by which to get along with others. He stated: "To go beyond is as wrong as to fall short" (p. 242). Based on these learning experiences, several organizations have been involved in implementing nonformal education and rural development activities in Thailand and Sri Lanka.

Most of Confucius's educational ideology was developed after he was thirty. As an adult educator, Confucius was one of the earliest recorded examples of someone involved in private teaching. As mentioned earlier, during his life, 3,000 disciples studied with him and 72 of them became distinguished. Many of his disciples were adults, and some were only a few years younger than him. Confucians, as a group, regarded education as the most important thing in life. Confucius said: "In teaching there should be no distinction of classes" (Legge, 1960, p. 305). This is one of Confucius's greatest educational thoughts. He advocated that people should have education whether they are rich or poor. However, this tends to be a contemporary interpretation of his philosophy. Galt (1951) argues that during Confucius's time, there was no conception of real democracy in China.

Confucius had a very serious attitude toward knowledge. Once, when he talked with one of his disciples, he said: "When you know a thing, to hold that you know it; and when you do not know a thing, to allow that you do not know it;—this is knowledge" (Pound, 1956, p. 151). His attitudes toward learning and teaching had no satiety or weariness.

Besides the basic educational theories, Confucius also suggested specific studying and teaching methods. He developed the method of elicitation and advocated a discussion method among his disciples and practiced this himself. His teaching was very often conducted in circles. There is no evidence that Confucius suggested the recitation method. In fact he encouraged people to understand the meaning of the context and put it to practical use. Confucius said:

> Though a man may be able to recite the three hundred odes, yet if, when entrusted with a governmental change, he knows not how to act, or if, when sent to any quarter on a mission, he cannot give his replies unassisted, not withstanding the extent of his learning, of what practical use is it. (Legge, 1960, p. 265)

When critics discuss traditional Chinese teaching methods, they often fault Confucius for using the "recitation" method. In fact, it is Confucius's followers who were at fault, having distorted his teaching methods and over-stressed memorization.

Confucius taught rites, music, archery, chariot, writing, and arithmetic. His curriculum content exhibited a kind of intermingling of moral and intellectual education (Zhu, 1992). Rites and music were the most important subjects according to Confucius. He also compiled his own textbooks, including the *Book of Poetry,* the *Book of History,* the *Book of Rites,* the *Book of Music,* the *Book of Changes,* and the *Spring and Autumn Annals.* Unfortunately, the *Book of Music* was lost. The other five books were revered as the Five Classics by later Confucian scholars and recited by thousands of scholars in preparation for the civil service examination.

Education after Confucius to 1911

Since the Han Dynasty, Confucian philosophy and educational ideology have played an essential role in shaping the Chinese education system. P. W. Kuo (1915), Dewey's first Chinese doctoral student at Columbia University Teachers College, pointed out that this high veneration for Confucius and the principles represented by him had an important bearing upon the subsequent history of Chinese education. From the Han Dynasty, Kuo argued, Chinese education became less liberal than it once was, and the content of education became narrowly confined to the Confucian classics. Kuo stated:

> From a larger point of view the over-emphasis upon the teaching of one school of thought to the exclusion of other systems originating during preceding dynasties, must be regarded as being extremely unfortunate for the progress of Chinese civilization. The study of Confucian classics

> became a habit of the student class who thenceforth held tenaciously to the saying of ancient sages and were afraid of advancing new thoughts of their own. They made no attempt to supersede the civilization of early antiquity and all they wished was not to fall too far away from it. As a result the thoughts of the scholar class continued to run in the beaten paths of the ancients, and no longer enjoyed the freedom necessary for all true advancement in civilization. (1915, p. 35)

What is particularly interesting in the evolution of Chinese education is not the detail of provision but its ethos and values. The primary goal of traditional Confucian education was to prepare the ruling elite and to mold the characters of its citizenry (Zhou, 1988). The whole process of learning was geared toward the memorization of ideas of antiquity, by way of the Four Books (*The Analects, The Great Learning, The Way of the Mean,* and *The Mencius,* compiled by Zhu Xia, neo-Confucian scholar in the Song Dynasty [960–1127]) and the Five Classics. This made up the content of education that had to be mastered especially for the civil service examination. Success in the examination was set as the end of learning (Cheng, Jin, & Gu, 1999). The sole reliance on classics excluded natural sciences and technical subjects from the curriculum. Furthermore, for centuries the emphasis in Chinese education was on rote-learning. This form of teaching tended to suppress the spirit of free inquiry and did not encourage any initiative on the part of students. It was assumed that the students would be submissive. The position of the learner was further compounded by the status and power of the teacher. In traditional China, teachers were listed among the five beings who should be most admired by society: the God of Heaven, the God of the Earth, the emperor, parents, and the teacher (Zhou, 1988).

Dewey and Progressivism in China

John Dewey may well be the singlemost influential philosopher of education the United States has produced, and his impact on all forms of education is immense (Elias & Merriam, 1995). However, his intimate association with recent Chinese history and education is amongst the most interesting but the least known (Su, 1995). From May 1919 to July 1921, Dewey traveled extensively in China and lectured on a variety of topics, including the goal of education, the relationship between school and society, moral education and democracy, and experiential learning. Among the many nations and regions he visited during his lifetime, Dewey spent more time in China than in any other foreign country.

Researchers claim that China was the foreign country in which Dewey exercised his greatest influence, particularly in the field of education (Clopton & Ou, 1973; Su, 1995; Xu, 1992).

This part will explore Dewey's visit to and his influence in China. There are three sections in this part. The first section provides a review of the major theories of Dewey's progressive education. Section two will explore his visit to China and the lectures he gave. The final section will provide an evaluation of Dewey's influence on Chinese education.

Major Theories of Progressive Education

John Dewey is regarded as the chief exponent of progressive education. Dewey (1916) believed that growth is the overriding aim of education and rejected the idea that education is preparation for work. Instead, he believed that schools should focus on the present lives of children. In *Democracy and Education* (1916), Dewey maintained that education has a role to play in social reform and reconstruction. In his view, education has both a conservative and a reconstructive function. He argued that education would flourish if it took place in a democracy; democracy would develop only if there were true education. For Dewey, a democratic society is committed to change and a democratic education would produce a society that is constantly in a state of greater growth and development.

Dewey (1916) defined education as the reconstruction and reorganization of experience that increases our ability to direct the course of subsequent experience. He (1938) argued that, in traditional education, students learn from texts and teachers; in progressive education, learning occurs through experience. Experience is the interaction of the individual with the environment through problem solving. According to Dewey, learning is not dictated by the teacher; rather, the teacher first attempts to help the student identify problems and then acts as a resource. Instead of using a teacher-centered approach, Dewey advocated a student-centered education.

Many people may see Dewey's theory of education as a theory of schooling for children. In fact, his ideas are also applicable to other forms of education and learning. Elias and Merriam (1995, p. 45) point out that "progressivism has had a greater impact upon the adult education movement in the United States than any other single school of thought." The authors also maintain that all of the major adult education theorists, including Knowles, Rogers, Houle, Tyler, Lindeman, Bergevin, and Freire, were each influenced by progressive thought.

Dewey's (1916) argument that education should be a lifetime commitment laid the basis for lifelong learning. He stated:

> Education must be reconceived, not as merely a preparation for maturity (whence our absurd idea that it should stop after adolescence) but as a continuous growth of the mind and a continuous illumination of life. In a sense, the school can give us only the instrumentalities of mental growth; the rest depends upon an absorption and interpretation of experience. Real education comes after we leave school and there is no reason why it should stop before death. (p. 25)

Dewey's Visit to China

Dewey visited China during a very significant period in Chinese history. The Opium War in 1840 marked the decay and decline of the feudal dynasty and heightened its social crisis. Many intellectuals began to turn toward the West for solutions and alternatives for reinvigorating the nation. They recognized the need to learn Western science and technology to reform the old system of education. In 1911, Dr. Sun Yat-sen led the Democratic Revolution, overthrew the rule of the Qing Dynasty, and established the Republic of China. Soon the democratic government lost its power to the warlords. There was still no democracy in China. This brought on the outbreak of the famous May Fourth Movement in 1919. It was, on the one hand, a nationwide student movement opposing Japanese imperialism and domestic Chinese corruption; on the other hand, it was a struggle between eastern and Western civilizations. It focused on whether China should adopt democracy and the sciences.

It is during these most critical years of modern Chinese history that Dewey, an established scholar in American educational philosophy and a Columbia University professor, was called upon by his former students in China to speak before the professors and students of the new Chinese universities (Keenan, 1977). According to Keenan, Dewey arrived in Shanghai on May 1, 1919, and stayed in China for a total of two years and two months. During this period, Dewey traveled to twelve of the twenty-two provinces and gave speeches in most of the cities he visited.

Dewey's Lectures

From 1919 to 1921, Dewey addressed Chinese audiences from some seventy-eight different lecture forums, and several of these were for a series of between fifteen and twenty lectures (Keenan, 1977). The major

Chinese journals and literary supplements throughout the country reprinted the Chinese versions of these lectures, and five anthologies were published. Nearly 100,000 copies of Dewey's principal series of lectures in Beijing (a 500-page book) were in circulation throughout China in 1921. Some continued to be reprinted for three more decades until the founding of the People's Republic of China. A number of Dewey's former students translated his lectures into Chinese. The translations and the publications in the Chinese journals were relatively accurate versions of what Dewey actually said in China (Keenan).

Dewey chose to base his lectures largely on three of his own books: *The School and Society* (1900), *Democracy and Education* (1916), and *Reconstruction in Philosophy* (1920). The lectures mainly came under three themes: modern science, democracy, and education (Keenan, 1977). Dewey linked the democratizing of society directly to the scientific revolution. His audiences in China were introduced to democracy and the philosophy of experimentalism in the same breath, with both portrayed as related developments in the history of Western thought. He attacked the notion of teachers passing knowledge on to students as if that knowledge were readymade and enshrined as permanent truth. In his China lectures, Dewey felt it important to emphasize the child-centered curriculum, a turning away from classroom emphasis on subject matter to emphasis on the growth of the child. He felt that child-centered education should be a priority for China as a departure from the stratified society or authoritarian tradition that tended to promote the "pouring in" of accepted subject matter as education.

There is no doubt that when in China Dewey delivered the message of science and democracy to his Chinese audiences. At the same time, he attacked the authority of the traditional Confucian education explicitly and implicitly (Clopton & Ou, 1973).

An Evaluation of Dewey's Influence

Though Dewey influenced China in many areas, this section intends to focus on his influence on education in China. While some scholars "praise him as a saint," others "condemn him as an enemy" (Su, 1995, p. 310). Reviewing historical development in China, the Chinese evaluation has gone through three stages.

The first stage covers the 1920s to the 1940s, when Dewey's pragmatic educational theory dominated the Chinese education field. His influence on Chinese education was intensive and prominent. Nearly all of his educational works were translated into Chinese, and some of them were used as textbooks in teacher education. His ideas were also adopted to transform the Chinese education system (Ou, 1970; Zhou, 1991). His former students and disciples (Chen Heqin, Hu Shi, Tao Xingzhi, and several others) in China played an important role in implementing his essential ideas. Many of them were already the intellectual leaders of the country.

The second stage was from the 1950s to the 1970s, after the new People's Republic was founded in 1949. This period was characterized by severe criticism and total denial of Deweyan experimentalism and his followers in China (Su, 1995). Dewey was portrayed as "anti-Marxist," "reactionary," a "defender of American imperialism," and "enemy of the Chinese people" (Chao, 1950; Chen, 1957). These criticisms were based on some of the arguments he made in his lectures in China and in his writings. One example was his opposition to the use of violence to overthrow the old system. While Marxists believed that communism would win the final victory in the world, Dewey maintained that the future was highly uncertain. Because of his emphasis on children's interests and experiences in the educational process, his educational ideas were criticized for lacking discipline, teacher authority, and rigorous teaching and learning in schools. All these qualified his theories as "poisonous and harmful" in China (Su, 1995).

The last stage started in the 1980s and continues to the present. During this time the political and philosophical climate in China has clearly shifted from Marxism to pragmatism. As the late Chinese leader Deng Xiaoping put it, "No matter if it is a white cat or black cat; as long as it catches a mouse, it is a good cat" (Su, 1995, p. 315). This indicates that if the economic theories and practices in the capitalist countries have resulted in a better living for the ordinary people, then there must be something of value for the Chinese people to learn from and apply to their own situation. Deng's political and economic pragmatism paved the way for Chinese intellectuals to become infatuated once again with Western pragmatism.

The new political situation saw the beginning of a serious reevaluation of Dewey's influence on Chinese education. Instead of totally denying Dewey, the Chinese critics realized that they should critically borrow and make use of Dewey's ideas in Chinese educational practice. Su (1995) argues that this period of re-evaluation is in many ways drastically different from the observation made in the 1950s. The new re-evaluation centers on the contributions that Dewey made to world education, the similarities between Dewey and Chinese educators and politicians, and the utility of Dewey's ideas for the improvement of China's educational practices. As Su notes, surely the Chinese educators will not totally abandon their established education system, but they now see the necessity to incorporate the useful elements from Western education, including Dewey's ideas, into the Chinese system.

A Comparison of Dewey and Mao

Many people in the world know Mao as a famous communist revolutionary leader. In fact, he was also one of the most important social and ideological forces in modern Chinese education. As mentioned earlier, in 1917 Mao started the first evening school for poor workers to raise their literacy in his hometown, Human (Xiao, 2003). Owing to his personal beliefs in the transformative roles of education and his supreme position in the Chinese Communist Party, its army, and its government, his thoughts on education extended to both theory development and policy making. Cleverley (1985) called him "the single most influential figure in the creation of a distinctively different communist system of education in China" (p. 70).

Similarities

Xu's (1992) comparative study of Dewey and Mao found amazing similarities as well as many differences between these two influential figures in modern Chinese history. Xu maintained that

> although Dewey and Mao emerged from completely different cultures, times, and contexts, their theories had amazing similarities. Their logos, "Learning by doing" (Dewey) and "Learning by practicing" (Mao) ring a similar note. Moreover, their views on the significant connections between school and society, the social role of education, the role of experience in learning, and their stress on moral education overlap a great deal. (p. 3)

Xu continued to argue that both Dewey and Mao believed that education was not an isolated enterprise but one "closely connected with, affected by, and achieved with and for social change" (p. 97). They both saw the necessity and significance of moral education in schooling, and placed it as the top priority before intellectual and physical development. Their notions of social characteristics of schooling and moral education led to an epistemological similarity. They both agreed that knowledge consists of experience and can only be acquired through active human inquiry in experience.

Differences

Xu (1992) also identified a number of differences between these two great thinkers. First, on the view pertaining to school and society, Dewey argued that school connected closely with society and played a vital role in social reforms and reconstruction. To Dewey, school is society. As a Marxist, Mao stressed that society is school. Mao maintained that, to make schooling serve for the proletarian state, a transformative education cannot occur within school itself but rather in the society at large. Second, Dewey's educational ideas were built on modern sciences and Western philosophy and pursued through academia, whereas Mao's were founded on Marxist political ideology and focused on social and political transformation via revolutionary struggle. Consequently, they clearly had quite different visions of what constituted moral education. For Dewey, moral education consisted of qualities such as democracy, open-mindedness, intelligence, intellectual honesty, and responsibility. Mao's moral education, on the other hand, had a strong political and class orientation, and demanded an absolute belief in Marxism and development of proletarian consciousness. According to Xu, their concepts of the most valuable experience in education also differed; natural sciences for personal and academic growth for Dewey, and political ideology for social welfare for Mao. Thus, the emphasis each placed on educational content was also different. Deweyan schools introduced modern subjects, a variety of topics, and experimental experience into their curricula, whereas Mao's schools substituted political study and productive labor as their main contents, in place of academic courses.

However, it is important to point out that both Dewey's and Mao's ideas challenged the traditional Confucian education and the old social orders, and served the progressive causes in social development (Xu, 1992). As mentioned earlier, Confucian education clung to classics as its only content and devalued any form of ordinary experience. Both Dewey and Mao advocated change and brought everyday experience into the classroom for educational purposes. Both strongly opposed rote-learning and imperial examination used by traditional Confucian educators. Instead, they guaranteed learners an active role in learning and took their interests into consideration. They both favored inductive methods, group discussions, and activities. They also focused on fostering imagination, originality, creativity, and the student's own capabilities of thinking and problem solving. Xu also noted that Mao's insistence that proletarian ideology was the only correct outlook as well as the only correct methodology cut away all the freedom and originality he advocated in education and, eventually and sadly, reduced his methodology to exactly what he set out to fight against—traditional cramming. Mao and his followers have also contributed to regression of education to rote-learning by cramming Marxism and Mao's thoughts into people's minds without any say from the people in general.

The above comparison demonstrates that, despite the fact that Dewey and Mao approached education from completely different cultural and ideological backgrounds, both of their efforts challenged the principles of the traditional Confucian education. To a certain degree, they removed modern Chinese education from the isolated ivory tower and situated it much closer within the current social reality.

Contemporary Education in China: Which Force Drives It?

The final part of this chapter will situate traditional Confucianism, Dewey's progressivism, and Maoism in the context of contemporary Chinese adult education and examine the roles played by each in shaping China's adult education, in the hope of determining which force(s) exert the strongest influence on China's current education system.

Education under Deng

The death of Mao and the arrest of the Gang of Four (Jiang Qing, Zhang Chunqiao, Wang Hongwei, and Yao Wenyuan) in 1976 marked the end of the Cultural Revolution and the beginning of a new era. Following the Third Plenary Session of the Eleventh Central Committee of the Communist Party of China in December 1978, Deng Xiaoping came to power. Deng's rehabilitation enabled him to implement his political and economic pragmatism in a number of new reforms, characterized by an "open door" policy and a socialist market economy. Very quickly, the nation shifted its emphasis from a "class struggle" to economic reconstruction. Modernization of agriculture, industry, national defense, and science and technology (known as the Four Modernizations) were identified as top priorities for the nation. Deng also reaffirmed the special role of education in the construction of socialist economy. He asserted at the National Conference on Education in 1978 that the realization of the Four Modernizations must rely on the development of science and technology, the foundation of which was education. He viewed education, particularly adult education, as an economically "productive force" (Zhou, 1988, p. 13). Teachers once again reclaimed their respectable status and were hailed as "glorious engineers cultivating human souls," and teaching became "the most glorious profession under the sun" (Li, 1999).

Within this context, education has undergone tremendous transformation under Deng's China. Many schools and universities that were shut down or moved to the countryside during the Cultural Revolution moved back to cities or began to offer academic classes. Various adult education programs, the use of radio and television, correspondence courses, and night universities the use of radio were provided to help "the delayed generation" to catch up with their education (Ke, 1992). Formal curricula with a focus on traditional academic subjects replaced those offered during the Cultural Revolution, which emphasized ideological and political study and physical labor. Meanwhile, secondary education expanded its vocational and technical education components in order to produce adequate manpower to support the country's economic development. After eleven years' abeyance, university entrance examination was restored in 1977 with a strong emphasis on students' examination results rather than on their political background. In a few years, "education was brought back to the highest point before the Cultural Revolution, but exceeded far beyond" (Xu, 1992). According to Fouts and Chan (1995), "The system began to resemble once again the Confucian model" (p. 527).

Renaissance of Confucianism in Deng's China

The previous analysis has revealed that, with the need for a skilled workforce to support the national agenda of modernization, economic construction, and market economy, China's education system in the last two decades has taken on "a more decidedly pragmatic and technological flavour" (Fouts & Chan, 1995, p. 528). While Dewey and Mao attacked many of the components of Confucian education, "more moderate later regimes left intact other elements to be used for various political purposes" (p. 527). Sciences, technology, Marxist ideology, Mao's thought, and Deng's theory replaced Confucian classics, but "the underlying Confucian traits of education (and society) remained basically unchanged" (p. 526). Owing to limited space, the following discussion will examine the manifestations of Confucianism in the areas of adult teaching and learning in China today, where its renaissance is most prominent. A number of studies (Guo, 1996; Hunter & Keehn, 1985; Paine, 1990; Pratt, 1992; Pratt, Kelly, & Wong, 1999) have demonstrated that present-day education in China is still teacher centered. Teaching and learning rely heavily on the use of textbooks, memorization, and examination.

Pratt (1992) and Pratt et al. (1999) report that teachers in China are regarded as content experts and transmitters of knowledge, while students are the consumers of knowledge. Teachers give and learners receive. Teachers are expected to be thoroughly prepared and organized for lectures. Paine (1990) illustrates this kind of teaching as the Chinese Virtuoso Model of Teaching, where the teacher resembles a musician who performs for the whole class and the students become an audience. The main intellectual thrust of this model of teaching, Paine continues to argue, centers on the teacher's performance and minimizes or inadvertently neglects the interactional potential of classroom experience. Teachers are still at the center of the stage. The Western notion of tailoring instructional processes to adult learner needs runs counter to deeply embedded Confucian beliefs about optimal teacher-learner relationships in China (Boshier, Huang, Song, & Song, 2006). Obviously, this process differs from Dewey's learner-centered education. The active role assigned to the learners by Mao during the Cultural Revolution is also missing from this process.

Learning is still seen as occurring through transference of information from the teacher to learner (Pratt, 1992). According to Hunter and Keehn (1985), learning in China is largely by rote, and teaching is by lecture with few aids other than the chalkboard. Except where learning is taking place directly related to a particular enterprise, there is no "learner involvement," no "participatory techniques," and no "problem-solving methodologies." Biggs (1997), on the other hand, cautions people that rote-learning is different from memorization. The difference between the two lies in the learner's intention with respect to meaning. In rote-learning, learners learn mechanically without understanding the meaning of the material, while memorization is a learning strategy and a repetitive way toward understanding the material. He maintains that Chinese students may be repetitive learners, but there is no evidence that they use rote-learning any more than their Western counterparts. Putting their differences aside, it is safe to conclude that neither rote-learning nor memorization is the kind of learning advocated by Dewey ("learning by doing") or Mao ("learning by practicing").

Knowledge is perceived by the Chinese as both external to the learner and stable in its movement from teacher to learner (Pratt, 1992). Usually there is little doubt about what constitutes the "basics" or foundational knowledge that students are expected to master (Pratt et al., 1999). The major source of that knowledge is usually authorized textbooks. As Paine (1990) puts it, "The textbook, as the source of knowledge, and the teacher, as the presenter of that knowledge, stand at center stage for the activity of Chinese schools" (p. 51). One of the criteria used to judge the effectiveness of teaching and learning is based on how well the teacher performs or transmits that knowledge and how well the student memorizes or masters it. It is also clear that this process of learning from texts contradicts Dewey's notion of learning as an individual's inquiry into students' experience. Furthermore, the kind of learning from the students' daily lives, political study, and physical labor advocated by Mao is also absent from this process.

The present-day education in China is still examination centered (Guo, 1996). Almost a century has passed since the civil service examination was abolished in 1905. The sciences, Marxism, Leninism, and Mao's thought have replaced the Confucian curriculum, but, in essence, the way of testing students remains much the same. While the content of examinations varies according to the subject and level, they all emphasize the testing of facts. Formal examinations only stopped for a decade during the Cultural Revolution, but resumed soon after it was over. Again, this is another legacy of Confucianism.

Who Is in Control?

The foregoing discussion has revealed that, despite Dewey's and Mao's attack on Confucianism, traditional Confucian traits still dominate teaching and learning in China today with the support of Deng's pragmatism. However, this does not mean that Dewey's progressivism and Maoism have completely vanished from China's present-day education. Fouts and Chan (1995) maintained that "elements of Mao's ideas were inculcated into the educational system" (p. 527). The authors also pointed out that Mao's universal education became the goal and that political education stayed in the curriculum along with moral education. The core of the five aims of education, that is, moral, intellectual, physical, aesthetic, and social, still remain as Mao's ideas. Following the June Fourth incident of 1989, the Chinese government under the leadership of Jiang Zeming adopted new policies of sending first-year university/college students to the army for a period of military training and of requiring two years' working experience before graduate admission. These policies can be traced to Mao's educational ideology and methodology. Although Mao is rarely quoted today, Xu (1992) predicts that his ideas have been and will be playing a significant role in the formation of China's contemporary and future education.

It is true that the influence of Dewey's progressivism on Chinese education was most profound from the early 1920s to the late 1940s. Like

Mao, his educational theories have not been completely shuffled into the museums. Dewey's views on the social role of education, moral education, vocational and technical education, and the connection between school and society have been partially incorporated into the Chinese education system (Su, 1995). As Cheng et al. (1999, p. 129) noted, "Education is acquiring a new meaning" in China. In the field of adult and lifelong education, they continue to argue, the focus has shifted from preparation or upgrading of manpower to personal development in the last few years. This new notion of education as a way of realizing a meaningful life resembles Dewey's view of education for personal growth. A similar recent move was the government's decision to promote lifelong learning in building a learning society (Boshier et al., 2006).

However, it is predicted that Dewey's ideas are not likely to become the focus of intellectual discourse in China as happened following the May Fourth Movement in 1919; nor will Mao's thought regain the omnipotence it apparently assumed during the Cultural Revolution (Tu, 1992). However, as Xu (1992, p. ix) pointed out, "Dewey and Mao are entwined theoretically and practically with the past, thus, they are inevitably embedded in the present, and integrated with the future." In fact, China is trying to avoid going to either extreme—"traditional education," as represented by Confucian educational theories, or "modern education," as represented by Dewey and his advocates in China (Su, 1995). It appears that, in the years to come, Dewey's progressivism and Maoism will join Deng's pragmatism, traditional Confucianism, and many other trends of thoughts in informing, defining, guiding, and challenging China's education system. Because all these theories and ideologies were developed in a completely different historical, social, political, economic, and cultural context, the implementation of each of them into practice in China inevitably requires interpretation, imagination, and adaptation. As Xu (1992) put it, "The implementation of Dewey would never be pure Deweyan" (p. 62). The same could be said about many other theories and ideologies as well. Each of them will evolve in a new social, political, and economic context. The development of each will bring changes to the education system in China.

In conclusion, this study reveals that China is contested ground for different ideologies. China's education system today is defined and guided by a synthesis of various social and ideological forces, consisting of traditional Confucianism, Deng's pragmatism, Maoism, and Dewey's progressivism. These forces, although often mutually exclusive, have formed a joint force together with other trends of thoughts. It is these joint forces that make the Chinese education system dynamic, unique, and fascinating.

Around the globe, educators are searching for answers to questions such as the relationship between school and society, the connection between theory and practice, and the choice between learner-centered and teacher-centered approaches. Many well-intentioned people may wish to experiment with implementing their favorite theory or model in a different environment. It is hoped that the history of Chinese education will provide a dynamic stage for educators in the world to reflect upon and learn from as they create and improve their own educational system.

References

A previous version of this chapter appeared in the *Canadian Journal of University Continuing Education, 30*(1), 2004. Thanks to the kind permission of the journal editor.

Biggs, J. (1997). Western misperceptions of the Confucian-heritage learning culture. In D. Watkins & J. Biggs (Eds.), *The Chinese learner: Cultural, psychological and contextual influences* (pp. 45–67). Hong Kong: University of Hong Kong.

Boshier, R., Huang, Y., Song, Q., & Song, L. (2006). Market socialism meets the lost generation: Motivational orientations of adult learners in Shanghai. *Adult Education Quarterly, 56*(3), 201–222.

Chao, F. (1950). Introduction to the criticism of John Dewey, Part I. *People's Education, 6,* 21–28.

Chen, J.P. (1957). *Criticism of Dewey's moral education philosophy.* Wuhan, China: Huber People's Press.

Cheng, K.M., Jin, S.H., & Gu, X.B. (1999). From training to education: Lifelong learning in China. *Comparative Education, 35*(2), 119–129.

Cleverley, J. (1985). *The schooling of China.* Boston: George Allen & Unwin.

Clopton, R., & Ou, T. (1973). *John Dewey: Lectures in China, 1919–1920* (An East-West Center Book). Honolulu: The University Press of Hawaii.

Dewey, J. (1900). *The school and society.* Chicago: The University of Chicago Press.

Dewey, J. (1916). *Democracy and education.* New York: Macmillan.

Dewey, J. (1938). *Experience and education.* New York: Collier Books.

Elias, J., & Merriam, S. (1995). *Philosophical foundations of adult education.* Malabar, FL: Krieger.

Fouts, J.T., & Chan, J.C.K. (1995). Confucius, Mao and modernization: Social studies education in the People's Republic of China. *Journal of Curriculum Studies, 27*(5), 523–543.

Galt, H.S. (1951). *A history of Chinese educational institutions.* London: Arthur Probsthain.

Guan, S.X. (1987). Developing a socialist adult education system in China. In C. Duke (Ed.), *Adult education: International perspectives from China* (pp. 196–200). London: Croom Helm.

Guo, S. (1996). Adult teaching and learning in China. *Convergence, 29*(1), 21–33.
Hunter, C., & Keehn, M.M. (1985). *Adult education in China.* London: Croom Helm.
Ke, M. (1992). The reform of adult higher certificate education. In X.D. Zhang & M. Stephens (Eds.), *University adult education in China* (pp. 141–145). Nottingham, UK: University of Nottingham.
Keenan, B. (1977). *The Dewey experiment in China.* Cambridge, MA: Harvard University Press.
Kuo, P.W. (1915). *The Chinese system of public education.* New York: Teachers College Press.
Legge, J. (1960). *The Chinese classics: Confucian analects* (Vol. 1). Hong Kong: Hong Kong University Press.
Li, D.F. (1999). Modernization and teacher education in China. *Teaching and Teacher Education, 15,* 179–192.
Ou, T.C. (1970). Dewey's lectures and influence in China. In J.A. Boydston (Ed.), *Guide to the works of John Dewey.* Barbondale & Edwardsville: Southern Illinois University Press.
Paine, L. (1990). The teacher as virtuoso: A Chinese model for teaching. *Teachers College Record, 92*(1), 49–81.
Pound, E. (1956). *Confucian analects.* London: Owen.
Pratt, D.D. (1992). Chinese conceptions of learning and teaching: A Westerner's attempt at understanding. *International Journal of Lifelong Education, 11*(4), 301–319.
Pratt, D.D., Kelly, M., & Wong, W.S.S. (1999). Chinese conceptions of "effective teaching" in Hong Kong: Towards culturally sensitive evaluation of teaching. *International Journal of Lifelong Education, 18*(4), 241–258.
Su, Z. (1995). A critical evaluation of John Dewey's influence on Chinese education. *American Journal of Education, 103*(3), 302–325.
Tu, W.M. (1992). Foreword. In D. Xu (Ed.), *A comparison of the educational ideas and practices of John Dewey and Mao Zedong in China: Is school society or society school?* San Francisco: Mellen Research University Press.
Xiao, J. (2003). Redefining adult education in an emerging economy: The example of Shenzhen, China. *International Review of Education, 49*(5), 487–508.
Xu, D. (1992). *A comparison of the educational ideas and practices of John Dewey and Mao Zedong in China: Is school society or society school?* San Francisco: Mellen Research University Press.
Yao, Z.D. (1987). Adult education theory and development in China. In C. Duke (Ed.), *Adult education: International perspectives from China* (pp. 13–18). London: Croom Helm.
Zhou, G.P. (1991). The spread and influence of modern western education theories in China. *Educational Sciences, East China Teachers' University, 3,* 77–96.
Zhou, N.Z. (1988). Historical contexts of educational reforms in present-day China. *Interchange, 19*(3–4), 8–18.
Zhu, W. (1992). Confucius and traditional Chinese education: An assessment. In R. Hayhoe (Ed.), *Educational and modernization: The Chinese experience* (pp. 3–22). Oxford: Pergamon Press.

CHAPTER 10

The Role of Continuing Education in Zimbabwe

Michael Kariwo

Introduction

Paulo Freire stressed in his writings that education is a political phenomenon that is not neutral. It is obvious that the educational process is frequently used by those who hold social and political power in order to maintain their position. However, it is also a truism that learning pays. Ball (1992, cited in Longworth & Davies, 1996) states that "training (at its best) will make nations and their citizens wealthier, societies more effective and content, individuals freer and more able to determine their lives in the way they choose" (p. 9).

The role of continuing education in an independent African country such as Zimbabwe is crucial. Zimbabwe became independent in April 1980. The disparities in education between the whites and the blacks were wide (see Table 10.1). The new government decided to widen access to education at all levels, including primary, secondary, and tertiary. The consequences of this policy decision were far-reaching in terms of financial and human resources. In the primary and secondary sectors the immediate problem was the shortage of qualified teachers. Enormous financial resources had to be allocated to education in order for the government to achieve its objective.

There is plenty of literature on adult education in Zimbabwe (Bennell, 1997, 2000; Chivore, 1998; Mudariki, 2003). Most studies have focused on adult literacy, with particular emphasis on women in rural communities. There is, however, very sparse literature on the current status of university continuing education. I attempt to explore the development and

Table 10.1 University of Zimbabwe enrolments

	1957	*1960*	*1970*	*1975*	*1980*	*1988*
Blacks	8	49	363	707	1,427	7,264
Whites	68	153	489	561	693	300
Others	-	5	85	93	121	135
Total	76	207	937	1,361	2,241	7,699

Source: University of Zimbabwe Registry.

role of university continuing education as a special branch of adult education, not withstanding other components of adult education, such as adult basic education/literacy, vocational education and training, and further education, or that continuing education is sometimes used interchangeably with adult education. Literature has generally been located in the area of adult basic education and training which is education for people aged fifteen and above who are not engaged in formal schooling. Continuing education is a part of adult education. In this chapter, I discuss some theoretical perspectives that underpin adult education in Zimbabwe. I also discuss the development of continuing education and highlight various political, social, and economic factors that have influenced the development of adult education in the country. I conclude by referring to some lessons learned from experience.

The Concept

According to the Hamburg Declaration on Adult Education, 1997,

> Adult education denotes the entire body of ongoing learning processes, formal or otherwise, whereby persons regarded as adult by the society to which they belong develop their abilities, enrich their knowledge and improve their technical or professional qualifications or turn them in a new direction to meet their own needs or those of their society. (UNESCO Institute for Education, 1997, p. 1)

As a generic term, *continuing education* refers to a program of study (award-bearing or not) beyond compulsory education. The Higher Education Funding Council of England ([HEFCE], 2003) defines continuing education as courses normally lasting less than one year and usually part time. Continuing education can be vocational or nonvocational. Millar (1991) uses a broad classification of adult education and continuing education defined as work related, specific, and aimed at increasing competence. Continuing education is, however, an elusive and contested concept as it takes shape within a specific context. In general, continuing education is the same as adult education at least in terms of being intended for adult learners. It does not normally include basic instruction such as literacy, English language skills, or programs such as vocational training. It is assumed that the student already has an education and is simply continuing it. Continuing education has grown to include education for licensing bodies. This is to encourage professionals to maintain their training and stay up-to-date on new developments. Conferences and seminars may also be designed to satisfy professional continuing education.

There are three institutions in Zimbabwe that have designated special units for continuing education, namely, the National University of Science and Technology (NUST), the Masvingo State University (MSU), and the Chinhoyi University of Technology (CUT). This is intended to widen access to education and at the same time generate income for these institutions. Continuing education is also offered at other universities in Zimbabwe. For example, the Women's University in Africa, which is privately funded, has taken on a new ethos in promoting the education of women at the tertiary level in order to address gender differences. In examining the concept across regions, it is apparent that there are variations in definition. In other regions, continuing education is synonymous with adult education and is offered to or undertaken by persons who have completed a cycle of full-time education during childhood.

Continuous learning is a term in use that is derived from the concept that education is not a once-in-a-lifetime experience. Each person needs specific opportunities to keep abreast with technological and social change. In modern societies, formal, systematic, and institutionalized continuing education plays an important role because it extends formal learning beyond youth. There is currently an information explosion, and technological changes are much more rapid than they were previously. Not only are people interested in upgrading their skills but they are also engaged in higher-degree learning. The term *continuing education* is often used interchangeably with *lifelong learning* although both are different. Lifelong learning is a strand of continuing education that includes formal and nonformal learning and normally continues through the life span of an individual. The elusiveness of the concept can be illustrated by university centers that have both continuing education and lifelong learning programs. Boshier (2001) states:

> Lifelong learning tends to render invisible any obligations to address social conditions. It is nested in ideology of vocationalism. Learning is for acquiring skills alleged to enable the learner to work harder, faster and smarter and thus enables their employer to better compete in the global economy. (p. 368)

In Zimbabwe, lifelong learning is far from an individual domain as espoused by Boshier. Rather, it is a national and political objective. One only needs to view the courses provided by universities under their centers of continuing education. In Zimbabwe, for example, these centers are located at the NUST, CUT, and the MSU. The first two centers were established primarily with income generation in mind. They offer courses on industry and commerce as part of skills upgrading. The third center was established to ensure adequate supply of trained teachers. At the same time, centers such as the NUST are responsible for running degree programs for full-paying students who are not able to access places in the government-subsidized (regular) classes.

The definition of continuing education embraces a wide range of courses from short-term certificate courses that last for a few hours to degrees completed over a prolonged period. Continuing education means providing opportunities for learning throughout life. Society is constantly changing, and the needs of individuals and organizations in adapting to change have to be met through formal and informal ways. In Zimbabwe, there are compelling factors that have influenced the development of continuing education, such as the limited access to education for blacks during the colonial period, globalization and technological change, and the need to fill the skills gap left following independence and by the mass exodus of whites. While continuing education in many countries is dominated by adult education, in Zimbabwe, it is both adults and youths who cannot gain access in the present formal system or in informal methods of education.

UNESCO defines adult education as

> the entire body of organized educational processes, whatever the content, level and method, whether formal or otherwise, whether they prolong or replace education in schools, colleges and universities as well as in apprenticeship, whether by persons regarded as adults (by the society to which they belong develop their ability.)

The initial role of continuing education in Zimbabwe was to upgrade unqualified teachers through a program called the Zimbabwe Integrated

Teachers Certificate (ZINTEC), which enabled teachers hired by the Ministry of Education to study and complete a teacher's certificate in a period of four years. Primary school enrolments increased from 1,235,815 in 1980 to 2,476,575 in 1995, an increase of 100.4 percent. The number of primary schools increased from 3,161 to 4,633 during the same period. The number of teachers rose from 28,500 in 1980 to 64,184 in 1995. Secondary school enrolments increased from 74,321 in 1980 to 711,094 in 1995, an increase of 857 percent. The number of secondary schools increased from 197 to 1,535, while the number of teachers rose from 3,736 to 27,320 (Ministry of Education, Sports and Culture, 1995). There has been a phenomenal increase in the student population.

Once this initial problem was under control, the government's next task was to increase access for those adults who needed to improve their qualifications, whether vocational or academic. There was a critical shortage of skilled manpower after the mass exodus of whites in the years after independence. This resulted in the expansion of the tertiary education sector. Other providers included the private sector and church organizations that saw the need to train people in particular vocational skills. After working for many years, adults feel the need to improve their skills and knowledge, partly because of changing technology but also in order to remain competitive in the labor market.

There is much overlap in the terms used in adult education, whether formal or informal. It is therefore necessary to define my perspective. With the establishment of new universities, a new development in continuing education has been the emphasis on centers of continuing education as a niche area. The centers have a double function, to widen access to education and to bring in revenue for the institutions.

Higher education in Zimbabwe generally covers institutions such as universities, polytechnics, and technical, agricultural, and teachers' colleges. The term *postsecondary* has a different meaning because students take public examinations at the end of four years and after the sixth year. Students can enter teachers' colleges and polytechnics after their four years of secondary education but are required to have at least two advanced-level passes (sixth year) to enroll at the universities. Before independence, and for the first decade after independence, most black students dropped out after completing the first two years of secondary education. They took a public examination, the Zimbabwe Junior Certificate, which qualified them for training as teachers.

Theoretical Perspectives

In this chapter, I apply theoretical lenses from human capital, social capital, and critical theories in the wider context of globalization in my discussion of the role of continuing education in Zimbabwe. Most governments have introduced education policies that are intended to drive the economy. Human capital theory is one basis for educational policies adopted in both developing and developed countries. This theory, in its recent form, stresses the significance of education and training as the key to participating in the new global economy. Formal education is valued as a private consumer good, a form of cultural capital (Bourdieu & Passerson, 1977) that allows those with the required qualifications to get ahead and stay ahead. In the absence of vibrant industrial labor markets, job prospects are poor, which places an inflated premium on educational credentials. Continuing education, whether by nonformal or formal methods, therefore plays an important role in improving the knowledge and skills of people.

The premise in social capital theory is that investment in social relations yields some positive returns (Bourdieu, 1986; Coleman, *1990*; Erickson, 1995). Individuals engage in interactions and networking in order to make profits. This concept is supported by an increased flow of information. Social ties located in strategic positions or hierarchies can provide an individual with useful information and opportunities, which would not be otherwise accessible. In social capital, there is a role played by one's social credentials and the influence of colleagues. I find this perspective convincing given the way individuals have moved up in the workplace and up the social ladder in Zimbabwe. I believe it is one of the factors accounting for the snowball effect in the enrolments at tertiary institutions in the country.

Critical theory in adult learning should have at its core an understanding of how adults identify ideology that is predominant in their thoughts and actions and in institutions of society. By emphasizing adult education in postindependence educational policies, African governments are sensitive to the hegemony that existed in the preindependence era. Feminism, on the other hand, argues for adult education for women who are disadvantaged when it comes to improving themselves educationally. Rural women in particular have responsibilities that prevent their attendance in evening classes or doing further studies part time.

Foucault (1980) believed that power was an escapable and ever-present force in human affairs. At the same time, Gramsci (1995) believed that there would always be hegemonic domination. At independence, African governments tried to replace capitalist hegemony with a working-class hegemony that represents the interests of the majority. Unfortunately, most independent African countries have failed in this respect because of poor accountability, lack of good governance, and corruption. Globalization is influencing developments in higher education the world over. As a concept, it is generally used to describe changes that are affecting daily life in all parts of the world. Adult education has played an important role in mitigating the impact of globalization in developing countries where there is concern for the majority of the poor people in rural areas or urban slums. Education and training are essential in preparing adults for the global market. In this respect, continuing education and lifelong learning have resulted in the widening of access to education.

The Context

The role of continuing education in Zimbabwe is better understood in light of the historical development of higher education in the country before and after independence. In a population consisting of 95 percent blacks and 5 percent whites, blacks had limited access to university education (see Table 10.1). The situation began to change toward independence in 1980. The African University was considered an anachronism and immediate steps were taken to correct what were perceived as anomalies. In some cases, this entailed changing legal instruments and administrative arrangements upon which the universities were founded.

In response to the challenge to transform the University of Zimbabwe, the then-vice chancellor stated: "I regard the University of Zimbabwe as first and foremost a developmental University, which is singularly animated and concerned, theoretically and practically with the search for solutions to the concrete problems of national Development" (Kamba, 1981).

Initially, efforts to change the curriculum were less successful. The intention though was to make the Zimbabwean university less of an ivory tower and more relevant to society. In this respect, adult and continuing education were to be complementary to other efforts.

Socioeconomic and Demographic Factors

Several factors have influenced adult education in Zimbabwe. One of these is the rapid population growth, which has been a major contributor to the pressure in terms of access to education. Zimbabwe has a population of 11.6 million, of which 51 percent are women. The growth rate of population is 1.1 percent. The growth rate for high-income countries is 0.3 percent, with Europe experiencing a negative growth rate of 1 percent. These statistics show that Zimbabwe has a young and rapidly growing population that has serious consequences for resources, which is evident in the difficulties that governments face in funding higher education.

The Zimbabwean economy, which is agro-based, has not been growing fast enough. In the 1980s, on an average the economy grew at around 3 to 4 percent and tended toward stagnation. In 1997, it grew by less than 1 percent. In the period from 1998 to 2000, there was negative growth. This was caused by a number of factors such as prolonged droughts, hyperinflation, and a critical shortage of foreign exchange. The country has made a number of attempts to turn the economy around starting with the Economic Structural Adjustment Program (ESAP) in 1991, but the economy has continued to decline. The government is thus finding it difficult to provide all the funding required for higher education and other sectors. Yet the demand for higher education has increased tremendously in the last few years. The government was silent regarding adult education policies in 1980, mainly due to the financial implications. However, the Williams Commission was very clear on the way forward. It recommended the establishment of a strong center of distance education at the University of Zimbabwe.

The Williams Commission of Inquiry report (1989) observed that continuing education is a special category of part-time education. Among older students there are many workers who wish to engage in university studies in order to update or fill in the gaps in their professional knowledge. This desire is also prompted by ongoing changes in technologies and the influence exerted by globalization. In 1988, at the University of Zimbabwe, 21 percent of the total number of students were part time. While this was a significant number, the new nation of Zimbabwe wanted more people to have access to higher education; hence the recommendation by the Williams Commission to establish a new public university and a strong distance education center at the University of Zimbabwe. This center later became the nucleus of the Zimbabwe Open University (ZOU). Today, there are eight public universities, all of them having some form of continuing education courses. There are three institutions, as already mentioned, with clearly defined units that undertake continuing education—the NUST, the MSU, and the CUT. However, the University of Zimbabwe continues to have the largest number of part-time students in the faculties of education and commerce.

Initially, part-time and continuing education students were predominant in the faculty of education. Today, the faculties of business and commerce, as well as engineering, have surpassed the numbers in the faculty of education mainly because of the need to raise funds within universities. In addition, the private sector supports workers by contributing to student fees as an educational incentive. However, not all students are sponsored by companies. Some pay for themselves in order to gain qualifications that would enable them to make a career change.

University Continuing Education in Zimbabwe

The emphasis on continuing education in Zimbabwe is driven by two factors, amongst others. First, there is a general demand for higher education that is not being met, even for students graduating from high school. The government has therefore put in place policies to increase access. Second, the universities have been struggling to increase enrolments with limited public funds. Continuing education has provided an avenue for generating income that can be used internally without much government control. This diversification of the revenue base has proven popular because it allows university administration the flexibility to allocate resources to the most pressing needs.

National University of Science and Technology, Bulawayo

The Center for Continuing Education at the NUST was established in August 2001. It was to be instrumental in achieving the university's primary mission—"the advancement of knowledge with special bias towards the diffusion and extension of science and technology through teaching pure research, applied research and the fostering of close ties with commerce and industry." The center has a mandate to facilitate and provide flexible continuing education and training opportunities through the various departments of the university.

The model adopted by the university is a mixed one. The center runs short-term courses for the private and public sector for people who are already employed. However, it also coordinates degree courses done on a part-time or modular basis, as well as full-time evening programs for students willing to pay tuition based on full cost. The center was running distance education programs using satellite and financed by the World Bank. These have now been discontinued after the assistance from the World Bank was terminated. The NUST Center for Continuing Education runs several courses, the most popular of which are in computing skills. The concept of parallel programs was introduced in order to widen access to students who were not able to gain admission to to the limited government-sponsored seats. The students pay full fees and receive instruction. The block release programs are part-time programs. Students spend blocks of time at the university before they go away to continue on a part-time basis. When one considers that continuing education and lifelong learning are one and the same, then parallel programs form an important component of lifelong learning.

Parallel and Block Release Programs

The university offers parallel degree programs in the following areas: computer science, operations research, architecture and quantity surveying, business management, banking, finance, accounting, journalism and media studies, library and information science, and industrial and manufacturing engineering. The following programs are offered on block release: Management Development Program, Diploma in Development and Disaster Management, Bachelor of Environmental Science Honors in Public Health, Post Graduate Diploma in Public Management, MBA (Public Management), and Bachelor of Science in Operations Research. All these programs are market driven. Full fees are charged. The ages of students range from nineteen to forty-eight. Continuing education students account for 30 percent of the total students in the university. The market is mainly government and quasi-government bodies, although some programs have been run specifically for the private sector. For example, the Diploma in Development and Disaster Management caters mostly to personnel. Females constitute about 50 percent of the total students enrolled at the Center for Continuing Education.

Chinhoyi University of Technology

The CUT was established in 1991. In 2004, student enrolment in undergraduate programs was 450. The students enrolled in continuing education constituted 30 percent of the total student population. The market for mature students is mainly government. There is demand in the technology areas, and CUT seems to be filling the gap in this area by offering certificate courses at the diploma level. The main programs offered are the Advanced Diploma in Creative Art and Design and the Advanced Diploma in Clothing and Textile Technology. Students can progress to the Bachelor of Science in Nursing Education, Bachelor of Technology Education, or Bachelor of Fine Arts.

Masvingo State University

The MSU is one of the new public universities in Zimbabwe. It was established in 2004. It has a department of adult and continuing education. Its first intake in August 2005 was small with five females and ten males admitted into the diploma program in adult education. The ages of the students range from twenty-eight to forty. The 2006 intake was just as small. Most students are from government as well as nongovernmental organizations such as Care Zimbabwe and the Legal Resources Foundation. The market is therefore mainly teachers, the police force, army, and employees of NGOs. The total student enrolment in the adult and continuing education department is at present only 2 percent of the total university population.

Women's University in Africa

The Women's University in Africa was established in 2004. It ushered in a new era in the education of women who are underrepresented in university enrolments, where they constitute less than 30 percent of total enrolments. In the sciences and medicine, the number of females is less than 10 percent. In its mission statement, the university states that it is dedicated to reducing gender disparity by providing a gender-sensitive and socially responsible educational training and research institution. This is a radical approach to university education in Zimbabwe, which has hitherto been male dominated. The new approach calls for new networks even at a global level. It is even a new understanding in the education of women. There are great intentions in the university's objectives to link poverty reduction themes and HIV/AIDS awareness programs to the education of women.

Distance Education in Zimbabwe

The Williams Commission (1989) recommended a strong center for distance education at the University of Zimbabwe. Distance education has been defined by the university as "an educational process in which the learner and the teacher are separated in space and/or time for much of the learning process" (1986). A much more persuasive argument to the government was the claim that distance education costs are around one-third of conventional face-to-face programs. This was going to be a more viable way of increasing access. The commission stated that distance education was to enable the old and young rural and urban people to have easy access to the means of learning, as well as to provide the country's workforce with opportunities to improve their skills. Women would also benefit from part-time evening or weekend programs as well as from distance education. Culturally they have different roles than men, making full-time attendance at university more difficult.

The initial establishment of a distance education department under the faculty of education at the University of Zimbabwe, which later became a fully fledged distance education institution in 1999, as the Zimbabwe Open University (ZOU), was a significant development in continuing education. The ZINTEC program used the distance education model in that students were in residence only for a very short period in the four years of their training.

Zimbabwe Open University

The ZOU is one of the key institutions in the country undertaking university continuing education. Enrolments in continuing education courses are 50 percent of the total university enrolments. The institution went through rapid growth since its establishment. The characteristics of the student population are shown in Table 10.2.

The ZOU was established by an Act of parliament, in 1999. It originated from the Center for Distance Education in the Faculty of Education at the University of Zimbabwe in 1993. In 1996, the Center for Distance Education became the University College of Distance Education. Three years later, on March 1, 1999, the college became the ZOU. It has the following ten centers:

Bulawayo Region, Harare Region, Manicaland Region, Mashonaland Central, Mashonaland East, Mashonaland West, Matabeleland North, Matabeleland South, Masvingo, and Midlands. There are three faculties awarding a range of degrees including the following:

- Faculty of Commerce and Law
- Bachelor of Commerce in Banking and Finance (4 years)

Table 10.2 Zimbabwe Open University enrolments

	1995	*1997*	*1999*	*2000*	*2001*	*2002*	*2003*
Males	2,166	8,857	9,679	10,240	11,432	8,149	11,824
Females	1,009	1,572	4,734	6,841	7,392	5,687	7,404
Total	3,175	10,429	14,413	17,081	18,824	13,836	19,228

Source: Ministry of Higher Education (2003).

- Bachelor of Commerce in Accounting (4 years)
- Bachelor of Commerce in Marketing (4 years)
- Bachelor of Commerce in Human Resource Management & Industrial Relations (4 years)
- Executive Diploma in Business Leadership (18 months)
- Master in Business Administration (21⁄2 years)
- Faculty of Arts, Education, and Humanities
- Bachelor of Arts in English & Communication Studies
- Master of Education in Educational Administration, Planning, and licy Studies
- Bachelor of Education in Educational Administration, Planning, and Policy Studies
- Bachelor of Arts Media Studies (4 years)
- Bachelor of Science Honors Degree in Counseling (4 years)
- Bachelor of Science Honors Degree in Psychology (4 years)
- Bachelor of Science Honors Degree in Special Education (4 years)
- Diploma in Education (Primary) (Matabeleland North & Mashonaland Central Region)
- Bachelor of Education (Secondary)
- Faculty of Science (4 years)
- Bachelor of Science in Nursing
- Bachelor of Science, Mathematics, and Statistics
- Bachelor of Science in Agricultural Management
- Bachelor of Science Honors in Geography and Environmental Studies
- Bachelor of Science in Physical Education and Sports
- Mathematics Bridging Course (1 Semester)

African Virtual University

In its attempt to expand distance education in Zimbabwe, the government has allowed externally sponsored programs such as the African Virtual University (AVU). One example is the use of educational programs beamed from Nairobi. This is a pilot project supported by the World Bank. It began with two sites, one at the University of Zimbabwe and the other at the NUST. It is a satellite-based distance education program transmitting video-based courses. The aim is to use technology to increase access to educational resources, particularly in developing countries. At both sites, AVU has had a tremendous head start due to World Bank funding. However, the NUST site is no longer operational due to foreign exchange shortages. Initial success was achieved under very difficult conditions of space and equipment and an unreliable power supply.

The importance of this mode of teaching is that it uses technology to provide some of the latest materials in tertiary education to students in developing countries. The problem is that connectivity is frequently broken because of poor telephone lines and power failure. There are also problems associated with certification and continuity with the rest of the national programs.

Conclusion

Zimbabwe has achieved great strides in its continuing education programs. This is partly due to the expansion of university education. However, other players such as church organizations, the private sector, and international donor organizations have made significant contributions. The need to train teachers posed a great challenge soon after independence in 1980, following the government policy to widen access at all levels of education. The government's strategy to use untrained teachers in the interim, who were to complete their certificates during their appointments, certainly paid off. This might have affected the quality of education but not substantially, as Zimbabwean education is still highly regarded in the international academic community. The new phenomenon of income generation in universities and the influence of globalization are driving continuing education in Zimbabwe today.

It is my view that the government needs to come up with another detailed study of tertiary education provision, as the only major commissioned document on higher education is more than twenty-five years old. It is recommended that the government should adopt a more democratic and transparent approach to educational policy review. The majority of policy decisions on higher education are made in the Cabinet. While the process at present appears democratic because parliament is composed of constituency representatives, there is no doubt that the final agenda is set in the cabinet. The introduction of white papers for consultation on new policies may be more transparent because the public is literate and should have an input.

References

Ball, C. (1992). *Profitable learning.* London: RSA.

Bennell, P. (1997). *Vocational education and training in Zimbabwe: The role of the private sector provision in the context of economic reform.* IDS Working Paper 74. Brighton: Institute of Development Studies.

Bennell, P. (2000). The impact of economic liberalization on private sector training provision in Zimbabwe. *Assessment in Education, 7* (3), 439–455.

Boshier, R. (2001). Lifelong learning as bungy jumping: In New Zealand what goes down doesn't always come up. *International Journal of Lifelong Education, 20*(5), 361–377.

Bourdieu, P. (1986). The forms of capital. In J.G. Richardson (Ed.), *Handbook of theory and research for sociology of education.* Westport, CT: Greenwood Press.

Bourdieu, P., & Passerson, J.C. (1977). *Reproduction in education, society, and culture.* Beverly Hills, CA: Sage.

Chivore, B.R.S. (1998). *Private tertiary education in Zimbabwe: Current and future trends.* Gweru, Zimbabwe: Mambo Press.

Coleman, J.S. (1990). Social capital in the creation of human capital. *American Journal of Sociology, 94,* 91–121.

Erickson, B.H. (1995, February). *Networks, success, and class structure: A total view.* Sunbelt Social Networks Conference, Charlestown, South Carolina.

Foucault, M. (1980). *Power/knowledge: Selected interviews and other writings, 1972–1977.* New York: Pantheon.

Gramsci, A. (1995). *Further selections from the prison notebooks: Antonio Gramsci* (D. Boothman, Ed.). Minneapolis: University of Minnesota Press.

Higher Education Funding Council of England. (2003). *About us: Glossary.* Retrieved August 24, 2007, from http://www.hefce.ac.uk/glossary.htm. Updated January 3, 2003.

Kamba, W.J. (1981). The university: From this time on. In N.T. Chideya, C.E.M. Chikomba, A.C.J. Pongweni, & L.C. Tsikirayi (Eds.), *The role of the university and its future in Zimbabwe.* Harare: Harare Publishing House.

Longworth, N., & Davies, W. (1996). *Lifelong learning.* London: Kogan Page.

Millar, C.J. (1991). *Adult education: Delineating the field.* Unpublished paper, National Education Policy Investigation, Johannesburg.

Ministry of Education, Sports and Culture. (1995). Harare: Government of Zimbabwe.

Ministry of Higher Education. (2003). Harare: Government of Zimbabwe.

Mudariki, T. (2003). *Adult basic and literacy education in Zimbabwe (2002).* International Conference on Adult Basic and Literacy Education in the SADC region, December 3–5, 2002. Pietmaritzburg: Center for Adult Education, University of Natal, pp. 204–228.

UNESCO Institute for Education. (1997). *The Hamburg Declaration on adult learning.* Hamburg: UNESCO.

University College of Distance Education. (1998). *Zimbabwe open university strategic plan.* Harare: University of Zimbabwe.

University of Zimbabwe. (1993). *A project proposal for the establishment of the Centre for Distance Education at the University of Zimbabwe.* Harare: Author.

Williams, P.R.C. (1989). *Commission of inquiry into the establishment of a second university or campus in Zimbabwe.* Harare: Government Printers.

CHAPTER 11

Popular Education and Organized Response to Gold Mining in Ghana

Valerie Kwaipun

Ghana's current gold rush has exploded at an unprecedented rate and magnitude. Yet, despite their enormous mineral wealth, local mining communities typify the "underdevelopment paradox" common to most extractive-based economies. This disparity has galvanized a collective response from local communities and their activist supporters. Multinational mining corporations have become increasingly influential actors in Ghana's economic scene, with wide-reaching consequences for local people's socioecological stability. Consequently, the plight of communities affected by mining is garnering more public attention and mobilizing communities toward increased resistance to these large-scale mining projects, alongside demands for better compensation and improved community infrastructural development.

This chapter will describe the response of local communities to recent increases in surface mining activity, specifically within the Western Region of Ghana, which endures the highest concentration of extraction. I will present a brief historical description of the political economy of Ghana's mining sector, followed by a discussion of the socioecological impacts of mining experienced in local communities. I will then explore the emergence of community responses, more specifically, the role of nongovernmental organization (NGO) community partnerships struggling to advocate for the rights of mining communities, and the role of adult popular education and learning in struggles pertaining to mining development–related displacements (Kapoor, 2006).

The Political Economy of Mining in Ghana

The clamor for gold in Ghana (formerly the Gold Coast) is not a novel phenomenon. In the mid nineteenth century, Europeans were insatiable in their quest for raw materials to fuel their burgeoning industrial economy, and as a result they established colonies to facilitate mercantilist trade systems. Despite a long history of indigenous gold mining practices in the West African region, British colonial interests had significant influence on what Agbesinyale (2003) describes as three distinct gold boom "epochs" or "Jungle Booms" in Ghana's gold production history. The first and second booms ended abruptly as British attention twice shifted to war in Europe, World War I, and World War II. The third and largest boom emerged in the late 1980s and continues on to this day.

In 1957, Ghana's independence ushered in a decidedly socialist, anti-colonial political agenda, and key sectors of the national economy, specifically mining, were reorganized to fall under state ownership and management. Despite political shifts to military rule years later, the nation maintained its state-centered mining policy and continued to shy away from foreign interventions. However, the effects of the 1973 OPEC oil crisis brought devastation to Ghana's national economy and, coupled with ongoing political instability, the mining sector experienced a state of distress.

It was in this uncertain economic climate that the government embarked on a Structural Adjustment Program (SAP) in an effort to secure loans from the World Bank and the International Monetary Fund (IMF) in 1983. Procurement of these loans is conditional on the borrower's adherence to a neoliberal rubric of socioeconomic "adjustments" prescribed by the institutions. Aside from fiscal policy changes (as in tariffs, taxation, and interest rates), SAP translates to a recession of the state's role in providing public services, such as healthcare, education, sanitation, and energy, in favor of privatization.

This new desperation for foreign direct investment (FDI) was critical in reshaping Ghana's mining sector, particularly the gold subsector. Consequently, with the adoption of the Minerals and Mining Law of 1986 (PNDCL 153), fiscal and environmental regulations in the mining sector were liberalized in an effort to entice foreign investors. These changes contributed significantly to unprecedented amounts of FDI flowing into the gold subsector. More than US$6.5 billion of FDI was poured into mining operations and related service companies from 1983 to 1999 alone (Agbesinyale, 2003, pp. 126–127), and more than 230 companies were registered with the Minerals Commission in 2001 as either prospecting or actively mining gold (p. 131)—six of which are among the largest multinational mining corporations in the world. As a result, production has increased more than 600 percent since 1985, with gold accounting for nearly 95 percent of total mineral exports by 1995 (Akabzaa, 2000, p. 13).

Despite the enormous amounts of gold produced by the sector over the last two decades, several extensive field studies agree that Ghana's mining industry remains a tight economic enclave controlled by a cartel of multinational mining corporations and question the actual impact of mining on Ghana's national economy (Agbesinyale, 2003, p. 208; Akabzaa, 2000, p. 20). In fact, it has been documented that Ghana itself earned only 5 percent of the total mineral export value, or US$46.7 million out of a staggering US$893.6 million in 2003 (United Nations Conference on Trade and Development [UNCTAD], 2005, p. 50), and that more than 80 percent of revenues are permitted to be held offshore (Agbesinyale, pp. 120, 208; Akabzaa, p. 20).

In 2003, the World Bank published a report describing their research on Ghana's mining sector. They calculated the industry's contribution to total government tax revenues in 2001 to be 4 percent, or about US$31 million—including US$10 million of income tax paid by mining industry workers and royalties of US$18 million, accounting for most of the mining sector's contribution to government revenues. The report states that generous tax allowances were to blame for such modest corporate income tax payments, despite their combined turnover in excess of US$600 million (World Bank, 2003, p. 13). The bank's report seems to support concerns that the cost of current mining practices in Ghana may outweigh their perceived benefits and recommends further research into the affected communities, concluding that:

> it is unclear what its true net benefits are to Ghana. Large-scale mining by foreign companies has a high import content and produces only modest amounts of net foreign exchange for Ghana after accounting for all its outflows. Similarly, its corporate tax payments are low, due to various fiscal incentives necessary to attract and retain foreign investors. Employment creation is also modest, given the highly capital intensive nature of modern surface mining techniques. Local communities affected by large-scale mining have seen little benefit to date in the form of improved infrastructure or service provision, because much of the rents from mining are used to finance recurrent, not capital expenditure. A broader cost-benefit analysis of large-scale mining that factors in social and environmental costs and includes consultations with the affected communities, needs to be undertaken before granting future production licenses. (World Bank, 2003, p. 35)

As foreign investment rushes into Ghana, mineral production and resultant profits are soaring. However, in the face of such intense mining activity, there is increasing resistance to current mining policies and practices, as local communities struggle with the widespread consequences of such rigorous operations.

Impact of Mining Development on Local Communities

The Western Region, particularly the Wassa West District that surrounds the urban hub of Tarkwa, is well known as the most heavily mineralized and mined area in Ghana and perhaps in all of Africa (Akabzaa, 2000, p. 31). Current surface mining operations pose a particular threat to nearby farm- and forest-based populations, as the majority of rural poor in Ghana derive their food and income directly from their lush natural surroundings. Due to its heavy rainfall, fertile soils, and rich biodiversity, the region enjoys almost year-round cultivation and has established itself as one of the most important farming regions in the country. In fact, farming is the cornerstone of the Ghanaian economy, with agriculture accounting for approximately 36 percent of Ghana's GDP. This is in contrast to only about 2 percent large-scale mining is said to contribute (Agbesinyale, 2003, p. 213; Akabzaa, 2000, p. 21).

The recent and controversial assent of the Minerals and Mining Act of Ghana (hereafter, MMA 2006) has generated contentious debate. While this new law does make strides in that it clearly affords rights to compensation and resettlement for landowners displaced by mining, critics point to numerous contradictions and inconsistencies in the document. For instance, while all Ghanaians are afforded the right to access the High Court to settle disputes under the national constitution, the MMA 2006 redirects all mining-related disagreements and matters of compensation to the office of the minister. It also affords the placement of any mineral, as well as any land or water in which it is found, in the hands of the president in trust for the people of Ghana. This clause allows the state to acquire land or authorize its occupation and use for purposes it deems in the public or national interest, which has been condemned by activists as fundamentally unconstitutional. While displaced landowners are legally entitled to fair resettlement and/or compensation under the MMA 2006, they often have no choice but to accept a mere fraction of the real land or crop value, rather than face an expensive and arduous legal battle to procure a fair settlement. Surprisingly, their situation is actually favorable to that of land users and migrant workers, who actually reside on the lands and farm there for subsistence. These land users are afforded no legal compensation in cases of mining-related land alienation and are often forced to seek out a living in overcrowded urban slum areas (Agbesinyale).

Environmental destruction caused by mining also has deep cultural implications, as evident in the evolution of a communal land ownership system rooted in the belief that land should be "held in trust not only for the living, but also on behalf of the spirits of the dead and the yet-to-be-born" (Agbesinyale, p. 203). It was believed that gold reserves would replenish themselves if gold is mined respectfully, and annual rituals to obtain the blessings of the traditional gods of the land are still performed on some of the gold mines in the district. These traditional belief systems have evolved over centuries and serve as an important indigenous method of natural resource conservation. However, as market-based economic forces penetrate these traditional enclaves, the sanctity of surrounding ecosystems and in turn an entire culture is threatened. Communal land ownership systems have historically offered security for the rural poor in Ghana. While these traditional systems do retain a degree of legitimacy, they are ultimately subordinated by national laws, namely, the MMA 2006.

Despite assurances from the Minerals Commission that local communities are given every opportunity to participate in the major processes and procedures related to the granting of mining licenses, communities and advocacy groups disagree. For instance, once environmental impact assessments (EIA) and other important regulatory documents outlining local ecological baselines and anticipated environmental and social impacts are made public, community members are permitted to come forward with any apprehensions within twenty-one days (Agbesinyale, 2003, p. 169). At first this protocol might seem reasonable, but the shortcomings and biases of this system are obvious when one considers the accessibility and distribution of these documents. EIAs and other information may be readily available in urban centers (as in Tarkwa and Accra), but they are not distributed amongst those living in the rural areas, who are directly affected. Further, these documents are written in a type of English "legalese," which the average community member does not realistically have the technical capacity to understand clearly. This is compounded by a community's naïveté regarding the detailed legal obligations of mining companies and the exact logistical role of regulatory bodies like the Environmental Protection Agency (EPA), the Minerals Commission, and the recently established District Environmental Management Committee (DEMC). As a result, large-scale mining projects generally lack the socioecological sensitivities that would be garnered from the active inclusion of these community perspectives.

Environmental Impacts of Mining

Surface mining technology is used by all multinational gold companies in Ghana. While celebrated for its cost-effectiveness, it is also extremely physically invasive and has many impacts on the surrounding environment. Surface mining is responsible for the significant deforestation of tropical rain forests, which were reduced from 8.2 million hectares in 1902 to only 750,000 hectares by 1997 (Akabzaa, 2000, p. 31). Although mining companies and the government clearly state their intentions to reclaim the damaged areas by replacing topsoil and replanting trees, environmentalists are skeptical and insist that such disturbance of the forest ecosystems is essentially irreversible. They warn that the introduction of new species during reclamation can upset existing ecosystems, resulting in a loss of soil fertility, the disturbance of delicate water sources, crop damage, and even changes in surrounding climate (Anane, 2003). Indirect deforestation also occurs when surface mining operations push small-scale miners and farmers out of traditional areas of operation and force them to work in "closed forest" reserves, causing significant damage. In addition, as available farmlands become scarce, the resulting quality of remaining lands suffers drastically, jeopardizing the productivity of the next growing cycle (Akabzaa & Darimani, 2001, p. 68).

While mining companies are bound by law to compensate farmers for crops lost to surface mining, the slower, more cumulative environmental damage to water and/or soil, as well as loss of land use, goes without reparation. Initial offers are often paltry considering the significant damage and inconveniences incurred. Further, unless one is willing to engage in legal negotiations, compensation is often inadequate to relocate and rebuild livelihoods elsewhere.

Substantial evidence has been collated concerning the detrimental effects of mining on local water systems in the Wassa West District, which is especially worrying as the Ghana Water Company Ltd. and rural water supply agencies estimate that 55 percent of the district's population depend solely on local rivers and streams for their drinking water (Agbesinyale, 2003, p. 267). Water contamination from mining operations is a daily concern for communities, illustrated recently when a major mining company was accused of intentionally disposing of their fecal waste into a major tributary, effectively poisoning drinking water downstream (Ghana News Agency, December 13, 2005).

The use of cyanide to process gold ore poses another significant threat. In spite of corporate reassurance about adequate containment facilities and emergency measures, more than four major cyanide spills have occurred in the Wassa West District in the last decade. While there have been no official deaths linked to the spills, a significant number of ill-health effects and the extensive destruction of crops and fish stocks have been reported (Agbesinyale, 2003, p. 291). Studies have also observed a disturbance of natural groundwater tables due to nearby mining excavations, leading to very low or no yield in local boreholes and hand-dug wells (Akabzaa, 2000, p. 66). Another study conducted in December 1998 by the Environmental Chemistry Division of Water Research Institute of the Council for Scientific and Industrial Research (CSIR) sampled local sources of drinking water in twelve communities in the Wassa West District and discovered dangerously elevated levels of pollutants, including fecal coliforms, suspended solids, manganese, iron, chromium, lead, and mercury, as well as acidity (Akabzaa, 2000, pp. 60–63).

However, the increased levels of mercury detected in local rivers and streams can most likely be attributed to illegal small-scale (*galamsey*) mining operations in the region, which commonly misuse mercury to process gold. Despite attempts to regulate small-scale gold mining in 1989 with the adoption of the Small-Scale Gold Mining Law (PNDCL 218), it is estimated that hundreds of thousands of these *galamsey* miners operate illegally (Agbesinyale, 2003; World Bank, 2003). *Galamsey* provides an important means of employment for many local people and attracts a number of workers alienated from their lands and traditional agrarian-based livelihoods due to surface mining concessions. However, the intensification of the *galamsey* industry in recent years has contributed significantly to land degradation and water contamination.

Agbesinyale's (2003) field study summarizes complaints from community members, describing how *galamsey* miners persistently degrade streams through bed excavations, panning, and the withdrawal of water for soil washings, thus causing stream diversions, siltation, erosion, and pollution (p. 265). The migratory and volatile nature of *galamsey* positions them as a serious ecological threat to surrounding areas, as they lack any enforceable environmental regulatory system. However, the respondents in the study also believe that *galamsey* miners lack the skills and resources required to repair and restore mined-out areas and call on the government and the mining companies to subsidize the training of miners in the area so they could implement environmental management programs (p. 279).

In spite of mounting evidence revealing a plethora of negative environmental impacts resulting from increased mining activity, the response from the regulatory bodies has been inadequate. The regulatory authorities such as the EPA and the DEMC are significantly disempowered due to inadequate staffing and a lack of resources, undermining their ability to enforce environmental quality standards. In fact, it is reported that the EPA-Tarkwa office employs only one program officer and two supporting staff to monitor the environmental behavior of more than eight multinational mining companies and legions of medium and small-scale operators in the district (Agbesinyale, 2003, p. 168).

Social Impacts of Mining in Ghana

Agbesinyale's (2003) extensive field research in the Wassa West District provides many insights into the impact of accelerated mining practices on the daily lives of the rural and urban poor who inhabit the communities surrounding mining concessions.

In Tarkwa, despite the hundreds of millions of dollars produced within the district each year from mining, more than 40 percent of the people earn less than a dollar a day, aptly illustrating what Agbesinyale calls the "gold-poverty paradox" (p. 208). Surveys conducted in the Wassa West District report that more than 51 percent of respondents feel that mining is a "curse" rather than a "blessing," indicating increased cost of living and widespread poverty, unemployment due to loss of farmland, exclusion of communities from benefits of mining, and the undermining of agriculture due to mining and air, water, and noise pollution as the most significant reasons for this conclusion (p. 309).

It is estimated that the recent gold boom has caused the populations of mining towns like Tarkwa and Prestea to double in the last thirty years, putting tremendous strain on urban infrastructure and facilities, with little done to ameliorate the changing economic and social demands in these areas (Agbesinyale, 2003, p. 219). Increased land alienation and the exclusion of youth from compensation and relocation packages are responsible for this growing trend of urbanization and the resulting social agitation amongst community members, who blame these young migrants for increasing incidences of overcrowding, loss of housing, increased drug use, and crime (p. 293).

Food security is also threatened, as substantial farmland is being swallowed by mining concessions. The attraction of youth to the *galamsey* camps has decreased vocational interest in agricultural production, effectively causing a "food paradox," as the nation's most fertile area finds itself heavily dependent on expensive imported foodstuffs (p. 232). This growing estrangement from local food production has increased food prices dramatically, with nonmining households in Tarkwa spending approximately 75 percent of their family income on food alone (p. 235). Relatedly, child labor and high school dropout rates are also high, as children work petty jobs for menial wages in an effort to contribute to household incomes. In addition, the relatively high salaries of mine employees inflate local prices, making housing and goods virtually unaffordable for nonmining households.

Another survey in this study focused on the specific impact of mining on women in Ghana and yielded many interesting observations and conclusions (Agbesinyale, 2003, p. 305). First, it was found that in urban areas women who engage in petty service industries may benefit from the gold rush initially. However, their complete dependence on the gold market means that a small slump in prices can quickly translate to economic devastation for them. Second, due to women's lack of land/resource control and their exclusion from major decision-making processes, their concerns around food security and the well-being of their children go largely overlooked, as most males are more concerned with the financial issues of land alienation, namely, compensation and royalty payments. Third, as the forests recede due to surface mining and galamsey activity, women are most acutely affected, through the exacerbation of "time poverty." Female respondents describe the laborious routine of having to trek great distances to reach their farms, fetch water, and collect firewood. In addition, it was reported that as options in farm/forest communities narrow due to land alienation and increased drudgery, women flock to *galamsey* camps and attempt to eek out an existence by selling food and alcohol to miners or by washing, pounding, and carrying ore. These *galamsey* camps are notoriously dangerous, causing dozens of deaths each year as rudimentary operations are prone to collapse and operators generally work without the most basic of safety measures (helmets, gloves, and boots) (p. 292). Many women are unable to cope with the bleak and hazardous conditions of these camps, pushing a substantial number of them to seek employment in the growing prostitution industry, despite the threat of HIV/AIDS (pp. 304–305).

The negative effect of mining on the health of local communities has been extensively documented (Agbesinyale, 2003; Akabzaa, 2000). In fact, the Wassa West District has the highest annual incidence of malaria in the country (almost five times the national average) due to the area's high rainfall and presence of large open pits from surface mines, which provide an ideal breeding ground for mosquitoes. Further, these field studies report a multitude of respiratory, skin, gastrointestinal, eye, and ear disorders and infections that have arisen as a result of intensified mining activity in local communities. In addition, the area suffers from a high rate of sexually transmitted diseases, including HIV/AIDS, as galamsey operators, local mine staff, and expatriate mine employees foster an expanding sex/prostitution industry.

The heavy social and environmental cost to communities due to recent mining operations has caused widespread exasperation and anger amongst residents and activists. As a result, a growing social movement has emerged throughout the Western Region, demanding increased corporate social responsibility, fair compensation, and accountability on all fronts.

NGO-Community Responses to Mining as Development and the Role of Learning

Current mining practices have met with considerable resistance from local communities, who feel increasingly excluded from the benefits afforded to mining corporations and their investors. Akabzaa (2000) explains that conflict and resistance in the mining areas generally revolve around issues of self-determination and resource control, distribution of mining revenue, land alienation, socioecological dislocation, and compensation/resettlement (p. 73). Growing frustration with these issues prompted the formation of a social-action NGO called the Wassa Association of Communities Affected by Mining (WACAM) in October 1998, which now consists of more than thirty-four communities and 20,000 members. WACAM seeks to network for the protection of the environment, natural resources and the rights of marginalized mining communities through advocacy, campaign, and representation within a legal framework that is sensitive to the concerns of mining communities (www.wacam.org). They are a registered and formalized NGO, yet remain very closely tied to popular movements struggling with mining in Ghana. In fact, the organization states that "WACAM should ultimately be transformed into a social movement well structured with resources and capacity to influence policies in favor of the marginalized people, especially those living in mining communities" (WACAM, 2005).

WACAM functions in a multitude of capacities, and volunteers can be found in courtrooms and boardrooms fighting for fair compensation and resettlement packages, attending and hosting international conferences focused on community struggles with large-scale mining, liaising with local and international NGOs on issues ranging from human rights to the environment, issuing press releases in response to various forms of corporate irresponsibility, and participating in a variety of community outreach projects or workshops. WACAM also serves as an invaluable support network for thousands of displaced people, at times organizing shelter, hospital care, and legal representation for those in need. WACAM-community partnerships have also been an important conduit for mining communities in Ghana to draw national/international attention to their struggle through both institutional (agency/government policy and practice, legal) and extrainstitutional (media, public opinion) channels. While the pursuit of institutionally based activism attempts to harness the power of the state to hold companies accountable for their actions, the use of media channels strives to raise public awareness and pressurize corporations to take the demands of mining communities more seriously by sensitizing consumers and hopefully affecting profits.

WACAM has widened its initial focus in the Wassa West District to include mining communities from across the country. While the organization does not officially represent all mining communities in Ghana, they do boast an impressive cross-national membership and have evolved into a national emblem of community resistance to mining development. Their active engagement with local civil society coalitions (National Coalition on Mining, or NCOM) and transnational advocacy networks/campaigns (Global Mining Campaign, GRUFIDES, Mining Watch Canada, and Oxfam) has helped to position them within a larger, global community of resistance. Yet, despite expanding their scope, the organization's refusal to compromise their mission and politics to procure support from funding agencies burdens them with significant financial constraints (allowing them to employ only four permanent staff as of 2005, several of whom maintained full-time employment outside of the organization to make ends meet). As a result, WACAM relies heavily on volunteerism from all members, whether urban-based intelligentsia or village farmers, to operate.

Community education is central to WACAM's mission and outreach/training programs to address a multitude of issues, such as the social/environmental impacts of mining and women's empowerment/leadership. However, most notably, WACAM emphasizes the importance of deepening the legal expertise of communities through paralegal education and employs a "rights-based" approach to advocacy. Through frequent collaborations with Ghana's Centre for Public Interest Law

(CEPIL), participants familiarize themselves with important legislative documents (such as the Constitution of the Republic of Ghana and the MMA 2006) and grow to understand current mining practices as a legal violation of human rights. WACAM then helps communities acquire the skills/expertise needed to engage appropriate judicial and bureaucratic channels. In addition to their focus on skills training and sensitization, these workshops also cultivate a more nuanced form of learning, which is deeply embedded within the participation of struggle itself and builds on Freire's (1970) conceptualization of conscientization—in other words, a process of learning that challenges and significantly alters participants' understanding of the world (Foley, 1999, p. 39), or the kind of learning that helps them understand their engagement in struggle as a transformative process toward the realization of social change (Kapoor, 2007). WACAM workshop attendees are often encouraged to reflect, discuss, ask questions, and draw on their common experiences. Such discussions show the beginnings of a collective unraveling of commonly held assumptions and the unmasking of contradictions within local perceptions of "reality."

For instance, in May 2003, WACAM began a series of workshops targeting assembly women and community leaders from mining communities in the Atuabu, Abekuase/Samahu, Damang, Bogoso, Nkwantakrom, and Ayensukrom zones. While these sessions were intended as a leadership training exercise, reports describe women engaging in extensive reflection-based activities around their role in society and in the local economy. Through this reflexive engagement, the women came to recognize their contributions as providers of unwaged labor and the importance of this work in maintaining existent, exploitative social structures (Owusu-Koranteng, 2003). Another example is taken from a project conducted by WACAM in February 2003, where participants from different communities surrounding the Bogoso Goldfields mining operations were asked to engage in a group exercise where they would compare the quality of their life before and after mining projects began in the area. Despite their acknowledgment of the company's acts of goodwill (donation of foodstuffs and drilling of several boreholes), there was an overwhelming agreement by all in attendance that their collective quality of life had deteriorated significantly due to mining activity.

> Our community was lively; people had money to care for every member of the family. Our children were in school and men were working. . . . Our community had nutritious foods and we were not getting sick often. We had money and the good things that nature could provide. . . . Our forest

> had Mahogany, and herbs. We were not getting sick frequently because we were having fish and meat for our meals from the forest. . . . Before surface mining, we had potable water from rivers Twigyaa, Monkoro, and Boadie. . . . Our whole lives and families are now disintegrated. We have no homes and live in great hardship. We were forced to abandon our farms and properties to relocate to other communities. We have become poor out of this. (WACAM, 2003)

These types of intercommunity gatherings facilitated by WACAM are also an important venue for exchanges between communities being prospected as potential mine sites to meet and learn from other communities who have lived with extensive mining activity for decades.

Outside of workshops and other explicitly educational forums, participation in the process of struggle affords valuable skill and knowledge gaining through the organization and execution of protests, petitions, rallies, community meetings, and demonstrations—where community activists deepen their organizational and administrative skills and understand that direct action is effective, flexible, and creative (Dykstra & Law, 1994; Foley, 1999). These events also act as platforms for the exchange of speeches, slogans, and symbols, which are themselves inherently educational and succinctly embody the spirit of the movement, helping to propagandize the goals and demands of the community toward the formation of a "resistance identity" (Routledge, 2003).

Fundamentally, movement actors have become increasingly skeptical of state/corporate promises and development rhetoric in general. Overall, they are more cognizant of underlying state/corporate agendas and, as a result, are more strategic in their dealings with such "players" (Kapoor, 2007). Through their partnerships with WACAM, communities challenge the assumptions of the development project and attempt to reclaim its meaning for their own benefit. However, WACAM-community partnerships are perhaps most successful in that they help to illuminate the linkages between resistance, action, and social change. This realization of agency is essential in a transformative learning process, as members become "subjects" of their struggle and learn that it is possible to acquire expertise, build new forms of organization, take action, and realize change. In light of, and perhaps to an extent, galvanized by, their encounters with obstacles throughout their participation in struggle, WACAM activists and community residents have realized relatively small but considerable achievements. While some of these achievements are explicit victories, others appear as quieter triumphs, realizations, and shifts in understanding.

Concluding Reflections

The globalization of industrial development, in the form of large-scale MNC-led projects, continues to meet resistance from those who depend on the sanctity of their surrounding environments to survive. Despite political freedom from colonial rule (see Kapoor in this collection), these subaltern (peasant/tribal/indigenous/women/farmers) groups continue to endure marginalization within their own sovereign societies, as they are forced to conform to the interests of national development and rapid industrialization and economic growth. As development's trajectory in the South follows Western paradigms of industrialization, progress, and modernity, the attendant dislocation and displacement of subalterns reiterates their exclusion from the national identity/conscience, as development is often conducted at their expense. However, their rejection of the wider development project does not necessarily imply a desire to "turn back time," or mean these groups are not interested in improving the quality of their life. Rather, they seek to reclaim the meaning and practice of development toward a perspective that privileges local, grassroots initiatives over global, grandiose action, promoting simpler, less materially intensive ways of living (Peet & Hartwick, 1999, p. 152).

Such struggles occur in the face of overarching, interrelated structural barriers (characteristic of subaltern realities in the South) that tilt the terrain of struggle to the disadvantage of local people. First, communities are consumed by daily struggles with poverty-related hardships, such as illiteracy, the exaggerated subordination of women, child labor, unemployment, threatened food security, and water scarcity, which are exacerbated by their dislocation from surrounding ecosystems and the social impacts of destructive development (as in large-scale mining). These issues make attempts to participate in a process of struggle that is much more difficult. Second, these movements are situated within a global capitalist economy, which prioritizes profit over community concerns around issues of health, education, and ecological sanctity. Structural adjustment and other neoliberal policies directly affect communities by forcing the state to endorse legislation conducive to increased FDI and a strong multinational presence, thus subordinating community demands for measures to temper socioecological devastation, which threaten to narrow profit margins and investor confidence.

However, important learning happens in confrontation with these barriers, because it is in the face of such hardships that communities convene, discuss, mobilize, study, observe, analyze, raise awareness, petition, protest, organize, plan, and learn in a process of struggle that is fundamentally educational. Through these incidental and inherently educative engagements with and through struggle, community activists are better able to politicize the process of development. Yet, despite this growing suspicion and ideological dismissal of mainstream development policy at the grass roots, in practice, it seems that community assertions tend to lean toward fundamentally reformist initiatives and continue to act within existing capitalist frameworks—preferring what Allman (1999) called a "reproductive" over "revolutionary" praxis. This is exemplified by a focus on more reactionary forms of advocacy, such as monetary compensation in the wake of environmental and social violations by corporate bodies. While mining communities undoubtedly deserve these reparations, there is a glaring lack of energy being invested into initiatives for their early inclusion in decision making nor is there an attempt to address community demands for resource control.

Therefore, social-action NGOs who partner movements need to recognize their unique position in opening up possibilities for new paths in struggle—paths that are decidedly political, historical, and collective. Given the multifaceted nature of learning within social movements, the acknowledgment and appreciation of the political nature of knowledge (more specifically, the recognition of relationships of power within its production and reproduction) are central to critical discussions of development and struggle. Recognizing the centrality of learning in shaping and directing struggle, particularly in the transformative potential of learning as conscientization, reveals the opportunity for a process of its repoliticization through what Freire (1970) refers to as a process of popular education (PE).[1] However, there is a need for activist-educators to also engage in a consistent process of self-reflexivity and remain aware of the "knowledge politics" they may introduce into a movement, despite their good intentions. Generally perceived in a position of heightened "expertise" and "authority" by communities, NGO activists need to scrutinize the sources of knowledge informing the struggle, as well as recognize the vulnerability of these partnerships to intra-struggle dynamics of power and their potential in creating new relationships of marginalization. As a struggle unfolds and activists become increasingly influential in illuminating possible directions for struggle, hierarchies are poised to develop and silently stratify a community. Therefore, it is imperative that activist-educators be aware of their influence within a community's understanding of struggle; the implications of these engagements on the direction of a movement, whose knowledge and/or ways of understanding are legitimized or in turn silenced through these partnered learning engagements; and whether or not they are sensitive to these voices.

Note

1. PE understands teaching/learning as an inherently political process, one where neither the teacher nor the learner is neutral. It seeks to address issues of oppression and struggle, recognizes the relationship between knowledge and power, and realizes the role of adult education/learning in confronting these oppressive relationships that typify subaltern struggles in the South. PE challenges an authoritarian model of education (banking education) and seeks to transform teaching/learning into a directed but democratic, participatory, dialogical, and problem-posing process—based on people's experiences, understanding, and knowledge—in an effort to transform the existent social scaffold of power through the persistent interrogation of the status quo and hegemonic assumptions that shape a person's perception of reality.

References

Agbesinyale, P. (2003). *Ghana's gold rush and regional development: The case of the Wassa West District* (SPRING Research Series No. 44). Germany: University of Dortmund.

Akabzaa, T. (2000). *Boom and dislocation: The environmental and social impacts of mining in the Wassa West District of Ghana.* Africa: The Third World Network.

Akabzaa, T., & Darimani, A. (2001). *Impacts of mining sector investment in Ghana: A study of the Tarkwa mining region.* Africa: The Third World Network.

Allman, P. (1999). *Revolutionary social transformation.* Westport, CT: Bergin & Garvey.

Anane, M. (2003). *Golden greed trouble looms over Ghana's forest reserves.* Retrieved February 20, 2006, from the World Rainforest Movement Web site: www.wrm.org .uy/countries/Ghana/Goldengreed.html

Dykstra, C., & Law, M. (1994). Popular social movements as educative forces. In M. Hayes (Ed.), *Proceedings of the Annual Adult Education Research Conference,* Knoxville University of Tennessee (pp.121–126).

Foley, G. (1999). *Learning in social action: A contribution to understanding informal education.* New York: Zed Books.

Freire, P. (1970). *Pedagogy of the oppressed.* New York: Continuum.

Ghana News Agency. (2005, December 13). *WACAM condemns disposal of faecal matter into Asuopre Stream.* Retrieved July 30, 2007, from http://www.ghanaweb.com/GhanaHomePage/NewsArchive/artikel.php?ID=96004

Kapoor, D. (2006). Popular education and Canadian engagements with social movement praxis in the South. In T. Fenwick, T. Nesbit, & B. Spencer (Eds.), *Contexts of adult education: Canadian perspectives* (pp. 239–249). Toronto, Canada: Thompson Educational Publishing.

Kapoor, D. (2007). Subaltern social movement learning and the decolonization of space in India. *International Education, 37*(1), 10–41.

Owusu-Koranteng, H. (2003, May 14–17). *Women leadership workshop for women in mining communities—Module One.* Workshop report. Tarkwa, Ghana: WACAM.

Peet, R., & Hartwick, E. (1999). *Theories of development.* New York: Guilford.
Routledge, P. (2003). Voices of the dammed: Discursive resistance amidst erasure in the Narmada Valley, India. *Political Geography, 22,* 243–270.
United Nations Conference on Trade and Development. (2005). Rethinking the role of FDI. *UNCTAD: Economic development in Africa.* (New York: Oxford University Press.)
Wassa Association of Communities Affected by Mining. (2003, February). *Baseline studies within communities affected by Bogoso Goldfields Limited.* Community meeting report.
Wassa Association of Communities Affected by Mining. (2005). Annual Report.
The World Bank. (2003). *Ghana—mining sector rehabilitation project and the mining sector development and environment project* (Report #26197).

CHAPTER 12

Popular Education, Hegemony, and Street Children in Brazil: Toward an Ethnographic Praxis

Samuel Veissière and Marcelo Diversi

Introduction

A chapter on street children may seem somewhat out-of-place in a work dedicated to "adult" education. Yet, as we hope to demonstrate, the phrase "adult education," despite its grounding in a critical tradition, indirectly refers to hegemonic cultural constructions of "childhood" and "adulthood," which in turn rely on dominant notions of "development." Such notions are necessarily problematized when one examines the lived experience of those human beings who, by their very mode of being, shatter conventional notions on the relationship between age—and development, behavior, place, and agency. The adult-child binary that separates legal agency from passive adult-imposed supervision in most official political landscapes is also reflected in formal education systems, which, as Paulo Freire famously argued, construct child learners as voiceless recipients of adult knowledge and culture. Critical pedagogy, conversely, recognizes the role of learners as cultural producers, active coinvestigators, and whole beings, as opposed to incomplete beings-in-the-making (Freire, 2000). The fact that such a critical paradigm has prevailed in many adult education initiatives while being almost systematically absent from formal schooling or any project dealing with children, thus, is undeniably a reflection of that legal binary opposition in which "adults" only are recognized as the owners of their existence.

Beyond the sadly obvious fact that street children subsist in that contested space where popular education initiatives are an urgent necessity, we would like to further justify their place in the realm of adult and popular education, and emphasize their capacity as active and resilient "whole" human beings who can draw on their resourcefulness and local knowledge to generate collaborative and emancipatory practices and become actors of meaningful social chance. Thus, we urge critical educators working with marginalized populations to look beyond hegemonic views on the relationship between age and agency, and consider children and adolescents as vital potential actors in collaborative processes of social change and emancipation that are informed by critical pedagogy.

In this chapter, we subsequently define our ideal paradigm of popular education as one that draws upon marginalized people's own resources and knowledge, while informing them about cultural and structural conditions of oppression and organizing them to take action (Kapoor, 2004). In contrast, our critique of the role of the broad range of nonformal educational structures (meaning, state or independent institutions that lie outside of formal public schooling) and nongovernmental organization (NGO) social initiatives, which we broadly define as "nonformal" and/or "popular" in the Brazilian context, seeks to expose the theoretical and pedagogical inadequacies of dominant approaches to "deal with the street children problem." There is therefore a difference—in quasi-Platonic tradition—between the "actual" popular education structures we are critiquing and the "ideal" model that we are proposing.

We show that dominant views of childhood are also linked to uncritical notions of social inclusion, and we argue that current popular education initiatives aimed at empowering street children lack the critical outlook to involve them actively in their emancipation and ultimately fail to understand the realities faced by these kids as well as the structural conditions that produced those realities. Drawing upon current anthropological literature and insights gathered from our own ethnographic fieldwork in southern Brazil (Diversi, 1998, 2003) and northeast Brazil (Veissière, 2007), we discuss the realities faced by street children, the failure of popular education programs to address these issues, and the role of such programs in maintaining and reproducing an oppressive status quo.

We conclude with suggestions on ways in which critical ethnography can inform popular education projects by providing alternative understanding of street kids' lived experience and situating their existence within broader hegemonic and structural conditions.

The Invention of Street Children: Nurtured and Nurturing Childhoods

Ever since the scandal around the assassination of eight homeless children—so-called *meninos de rua*—by a small group of off-duty police officers while the kids were sleeping in a huddle near Candelaria Church in Rio de Janeiro on July 23, 1993, the existence and harsh living conditions of Brazilian "street children" have come to the attention of the world's media, and exposed a sizeable thorn on the side of Brazil's carnivalized celebration of its modern democracy. While the Candelaria massacre triggered local and international protests by child rights advocates and uncovered further scandals involving organized police and paramilitary death squads, it also emerged that a significant part of the Brazilian public sided with the police and militias and called for the control—and, when privately questioned, extermination—of this "criminal" street "vermin" (Diversi, 1998; Goldstein, 2003; Hecht, 1998; Scheper-Hughes & Hoffman, 1998; Veissière, 2007). If the human rights activists and these "assertive" members of the Brazilian "public" do not seem to agree on the outcome of "what is to be done about street-children," they would nonetheless concur that the street is indeed "no place for children," and are in fact all eager to see these small human beings reintegrate into some sort of a status quo. In a hypothetical discussion between an activist and a conservative "concerned citizen," there would be no doubt considerable disagreement about the extent to which the structural conditions that produce situations where children come to live and work in the streets have to be altered. Where the argument would end, however, is on the consensus that unsupervised children do not belong in the streets, and on the fact that the indeterminate, unrestrained, "sexual," and "criminal" living conditions of *meninos de rua* are not appropriate for children. From this agreement, we begin to form the idea that street children are always perceived as a violently incongruous and "disorderly" vision, because they live not only outside the spatial and social boundaries where the laws of socioeconomic segregation would have relegated them but also outside the very discursive boundaries of what most of us have come to understand as "childhood." In his compelling ethnography of children's lives on the streets of Recife, Tobias Hecht (1998) raised important questions about culturally constructed categories such as "childhood" and "adulthood" and pointed out that the general consensus that street kids have been "robbed of their childhood" sheds light on what is meant by childhood in the Western liberal tradition. By their very existence, argued Hecht, street kids destabilize conventional notions of the relationships between age and behavior. While, as he reasoned, we might all wish for better living conditions for children who live on the streets, we are forced to acknowledge that there are indeed different ways of expressing and living one's childhood. Expanding on the concept of childhood as a discursively constituted category, Hecht critiqued other dimensions of this childhood consensus in which children are typically seen as vulnerable, passive, and incomplete beings-in-the-making whose sole purpose is to be the happy and docile recipients of adult culture. While Hecht did situate himself as one who ultimately wished to "rescue" street children, he also confessed ample admiration for their resilience, insisting that it provided a strong reminder that children are also whole beings and cultural producers who are active shapers of their cultural environment. The fact that children are traditionally seen as passive and voiceless has been widely critiqued in the interdisciplinary study of childhood and has prompted some scholars in the anthropology of childhood to describe children as colonized populations. As the American ethnographer Herb Childress noted,

> Developmental terminology is often dangerous, because it presumes that some people—that is, the definers of the terminology—are developed and have reached some pinnacle of being that all other groups are both preparatory to and desirous of. Like earlier non-modern groups and the present day equatorial and southern hemisphere worlds, which we often call "pre-industrialized nations", [children and] teenagers are seen as pre-adult, which means that . . . they are about to be colonized by a greater power, and that they should appreciate the intervention. (Childress, 2004, p. 196)

Hecht also remarked that there was no place for autonomy in dominant "colonizing" views of childhood, and located the main distinction between "normal" or "home" childhoods and those that were "robbed of a childhood" around the idea of nurture. "Normal" children, argued Hecht, were essentially "nurtured" beings, who were clothed, fed, loved, and cared for by their parents, families, and other cultural, social, and economic structures. Conversely, "nurturing" children—who, in urban contexts, are seen roaming, working, begging, or "living" on the streets—have to fend for themselves and often contribute their meager earnings to feed their families. Several ethnographic studies found that a majority of those who are traditionally described as "street children" return home at night after working in the streets, while a few alternate between sleeping at home and in the streets, and an even smaller group resides in the streets permanently (Diversi, 1998; Hecht, 1998; Scheper-

Hughes & Hoffman, 1998). The fact that children often migrate between these categories makes it difficult to quantify the number of truly homeless children. Whatever the numerical consensus and spatial practices, street kids represent a small, but most visible, portion of the disenfranchised Brazilian population. Yet, the crucial issue in this politics of definition, as Nancy Scheper-Hughes and Daniel Hoffman pointed out, is that the insistence on calling working, nurturing children "street kids" and the frustration over their visible existence reflect the preoccupations of one class and segment of Brazilian population with what they wishfully perceive as the "proper place" of another. As Scheper-Hughes and Hoffman aptly put it,

> Mary Douglas's definition of "dirt" as perfectly ordinary soil that is out of place comes to mind in this regard. Soil in the ground is clean, a potential garden; soil under the fingernails is dirt, a potential contaminant. Similarly, a poor, ragged child running unsupervised along an unpaved road in a *favela* or playing in a field of sugarcane is just a "kid", an unmarked *menino* or *menina*. That same child transposed to the main streets and plazas of town, however, can be seen as a threat or a social problem: a potentially dangerous (or potentially neglected) *menino de rua*, a "street kid." (p. 357)

The Tip of the Iceberg: The Structural Production of Nurturing Childhoods

The presence of autonomous children in the streets, as we have seen, taunts affluent Brazilians with a painful reminder that the exploitative system in which they enjoy privilege has reached a cancerous phase. This "reminder," however, is to them the only nagging visible part that lies at the fringe of a gigantic sphere of human misery produced by five centuries of colonial and neocolonial oppression. This neocolonial disaster, as many critics argued, is the product of an oppressive racialized class system in which the affluent have always sought to exploit and control the poor, while paradoxically seeking to shield themselves from their "contaminating" presence and concealing uncrossable social fractures through the dissemination of a deceiving "racial hegemony" (Hanchard, 1999) that perpetuates the false notion of a Brazilian racial democracy (Goldstein, 2003; Hanchard, 1999; Veissière, 2006, 2007).

Since the 1980s, equally exploitative trends in the global economy have significantly worsened conditions of employments for the poor while reducing the quantity and quality of state involvement in the public infrastructural sphere. In light of the recent neoliberal Structural Adjustment programs (SAPs) imposed by international financial institutions like the International Monetary Fund (IMF), the World Bank, and the Inter-American Development Bank (IDB), education systems and other health and social safety nets rapidly disintegrated and were forcibly decentralized, privatized, and thrown into the "free lay of the marketplace" (Arnove, Franz, Mollis, & Torres, 2003; Torres, Arnove, Franz, & Morse, 1997). Although such policies were designed to reduce Brazil's fiscal deficits and bring inflation under control, they contributed to increasing poverty and a widening of the gap between rich and poor while benefiting a small elite (Arnove et al., 2003).

Research data from the Economic Commission for Latin America and the Caribbean (ECLAC/CEPAL) showed that in the 1980s and early 1990s, 25 percent of the poorest households in the metropolitan areas of Rio de Janeiro and São Paulo lost 15 percent of their income, while 5 percent of the richest saw theirs increased by approximately 25 percent. Furthermore, the commission revealed that such losses were not only experienced by the poorest, and that as much as 50 percent of families located in the "middle" lost between 3 and 10 percent of their income (CEPAL, 1991, cited in Torres et al., 1997).

Extended family structures, which had traditionally played a strong nurturing role in Brazilian public life, underwent a similar disintegration as men and women were forced to migrate regularly in search of work. Thus, "typical" working-class family structures increasingly became matrifocal, with a single, toiling mother as the head of a household in a favela. Children of such households typically become nurturing children, because the absence of state and family support forces them to become autonomous and to become nurturers for other members of the household (Diversi, 1998; Hecht, 1998; Moulin & Pereira, 2000). Moreover, what little "public" state involvement subsists in the form of schooling is not only structurally inadequate because of lack of funding but also culturally and economically inadequate for favoring an alienating Eurocentric cultural capital, failing to address the living conditions of favela children, and failing to provide them with immediate and long-term resources for feeding themselves and their families and overcoming their situation of poverty (Arnove et al., 2003). Indeed, as Torres et al. (1997) demonstrated, Brazilian and other Latin American elites have successfully managed to maintain people of their social groups in positions of power by underfunding public primary education and investing in "public" universities that are accessible only to the wealthy who received a private primary and secondary education. We should point out, however, that Brazilian educational policy and practice cannot be said to have always been underpinned by such an ominously simplistic oppressive logic, but it has alas been disintegrated almost beyond repair by the totalitarian modernism of the military dictatorships of the 1960s and 1970s and the transition into late capitalism. While Brazil may still pride itself on a generation of professionals and intellectuals who were able to attend free universities after undergoing public schooling in the 1950s, the same cannot be said of younger generations who were halted by the deterioration of the country's elementary and secondary public systems in the 1960s and 1970s engineered by the modernists' avid pursuit of industrialization without educating the labor force.

In the contemporary socioeconomic landscape, thus, even if one were to leave aside the otherwise decisive questions of access—physical distance to schools and poor public transportation comes to mind—the paralyzing effects of hunger and poor nutrition on concentration, and the utterly unrealistic expectation that nurturing children can find the time and energy to dedicate themselves to full-time schooling, leave us with the sad reality that street and favela children would still find themselves doubly excluded by a public school system that is not only culturally and vocationally inadequate but also structurally unviable.

Nonformal/Popular Education and NGOs as Agents of Hegemony: Insights from Ethnography

As the presence of children in the streets is universally recognized as "a problem," many state, nongovernmental, and religious organizations have been put in place to offer them alternatives, such as shelters, drug rehabilitation centers, and the infamous reform schools of the State Foundation for the Well Being of Minors (FEBEM). Indeed, Tobias Hecht calculated that from mid-1992 to mid-1993, there was approximately one adult working for each child sleeping in the streets of Recife (1998, p. 23). This approximation, we should point out, does not amount to one educator for every working child.

All ethnographic studies that investigated the interaction of street children with these organizations, particularly the FEBEM reform schools, however, found that they most often only served to send these children back into the silent status quo of poverty, or even reinforce the status of "criminal" imposed on them by the affluent society (Diversi, 1998; Hecht, 1998; Scheper-Hughes & Hoffman, 1998). This phenomenon echoes Erving Goffman's pioneering study of asylums in which he demonstrated that mental institutions do not "cure" patients of their "illnesses" but ultimately inscribe the illness onto them by implicitly teaching them how to assume the role and display the symptoms of deviance. Such "total institutions," argued Goffman, typically produce the illnesses they are intended to treat (1961). Accordingly, FEBEM reform schools in Brazil have traditionally produced, or significantly contributed to, the marginality of street children by incarcerating them in institutions where they effectively learn to become what they are accused of being (Diversi, 1998; Hecht, 1998). Street kids' narratives reveal that institutions created for the "well being" of "minors" have characteristically apprehended individuals who were forced to work in the streets in order to survive, and punished and incarcerated them for not being the right kind of minor. Thus, street children who were imprisoned in reform schools consistently recount how they learned to become violent, "tough," and cunning as a response to the violence inflicted onto them and the systematic expectation from all authority figures that they could not be trusted (Diversi, 1998; Hecht, 1998).

Despite Lisa Markowitz's call for the ethnographic investigation of NGOs, and of the political sphere of interaction between international, institutional, and local "recipient" actors (Markowitz, 2001), there is at present little known work on the relationship between street children, NGOs, and social change, other than that written by NGO staff (Markowitz). The few examples that can be drawn from current literature dealing with street children and NGOs in Brazil indeed seem to suggest that such organizations operate more within the realm of social reproduction than that of social change, though not all agree on the extent to which they inscribe criminality onto "beneficiaries." Tobias Hecht (1998) found that while FEBEM reform schools tend to reinforce the status quo of street children's marginality, other NGOs typically aim at reinserting them into the cultural status quo of nurtured childhoods, which alas often turns out to be matrifocal households of the favelas where children are forced back into nurturing positions. Hecht pointed out that some UNICEF-sponsored programs and other international NGOs (such as Childhope) dealing with street children in Brazil have received considerable media attention. He argued, however, that such media coverage is generally intended as a way to raise funds for these institutions while little is being said on their impact on society and the way the children perceive them. Hecht's exegesis of this thorny issue is most pertinent to our argument because it enables us to look beyond the Goffmanian cliché of authoritarian total institutions and identify more subtle and ambiguous ways in which NGOs come to operate within the status quo. Hecht observed that these institutions' motives for "rescuing" street children are habitually situated within narratives of "salvation," "reclamation," or "citizenship" (1998, p. 23), while the kids characteristically come to view these social agencies as "an integral part of street life" but not as a "way out" (p. 24). Spending time with street children and discussing their relationship with NGOs, thus, he found that the street children of Recife often referred to these institutions as *fregu̇ẽses*—which he roughly translated as "clients"—meaning one of several alternatives to stealing and begging in order to secure food, clothing, and money. As one youth explained to Hecht about his resilient approach to survival, "I beg, panhandle, steal, whatever falls in the net is fish [*o que cair na rede é peixe*]" (p. 180). Hecht concluded that, to a large extent, NGOs and other nonformal social institutions are viewed by street children as "one type of fish that falls in the nets." In a similar vein, Robert Cubillos (2002), who investigated the streets of Recife after Tobias Hecht's insights, argued that given the current socioeconomic landscape, street children will continue to inhabit the streets and will keep using the "resources" provided by NGOs in order to "secure their survival and success" in the streets.

During his fieldwork carried out over several years with the street children of Campinas in the state of São Paulo, Marcelo Diversi (1998) encountered similar cases of nonformal state institutions that reinforced the children's marginality, and salvation-driven NGOs with superficial understanding of structural constraints and the kids' lived experience with which the children had ambiguous relationships. He came to realize that many of the street youth often utilized these institutions as temporary resources or shelters, and also that those who entered programs of institutional care in the hope of "curing themselves" from street life typically received confirmation that they were "not fit for normal life" as a result of their interaction with the institutions and their personnel, and found themselves violently ushered back into the status quo with their hopes shattered.

The well-intentioned but patronizing philosophy underlying all service programs made available to the street children of Campinas failed to account for the children's own perspectives of their condition. NGOs offering shelter failed to realize that children preferred the unsupervised and unstructured "shelters" of the streets. Programs offering educational services failed to understand that the street children they were trying to help did not find meaning in developing a trade—most children felt they were getting by just fine with their street smarts. Programs offering health education about nutrition, AIDS, STDs, and personal hygiene lost the children's interest from the beginning—for most street children did not think they would live long enough to have to worry about the future.

The youth Diversi worked with liked the street life, preferring the freedom of unsupervised life to broken homes and NGO programs born out of middle-class values. While most street children expressed gratitude for the services provided to them, their lack of interest in the offered services showed that, for them, the street was still the best alternative. Once again, here is the intersection where critical ethnographers can bring the most significant contribution: by assessing street children's perspectives of their lived experiences, critical ethnographers have the unique opportunity to inform and shape programs that offer alternatives more enticing than street life.

During his research in Salvador da Bahia, Samuel Veissière found that the street children he befriended often did not differentiate between state-run and nongovernmental organizations, which they all seemed to associate with a crumbling, corrupt, violent, and misguided "public system." There was, in the children's accounts of their relationship with "the system" and in what Veissière observed during his year of fieldwork in Bahia, a mixture of cynicism (with near-affectionate attitudes toward street educators' "cluelessness" and naïve proselytism), client-based resource-seeking behavior (like the instances documented by Hecht and Cubillos), and fear of beatings and imprisonment due to perceived relationships between the aid/charity sector and the police. From what Veissière witnessed in Salvador, the only street educators who were successful in developing meaningful and transformative relationships with street children were those who shied away from proselytism and the usual obsession with "reintegration" and getting children "off the streets," and who instead embarked upon the humbling and enriching experience of learning from the children's acts of resistance. One local NGO, which prided itself on a Freirean tradition of critical education going back to an actual collaboration with Paulo Freire, instructed its members to value, respect, and learn from what one of its coordinators labeled "the Call of the Street." "Our pedagogy," he had told Veissière,

> is one of listening. We spend a lot of time with kids in their environment, we do a lot of listening, and we don't do a lot of talking. Only later, much later, when the kids start talking about what took them to the street, do we start discussing things in more depth; then we can start to talk about race, to talk about class, to talk about history, about colonialism, and to situate their struggle historically. (Veissière, 2007, p. 46)

Though they did not call themselves ethnographers, the educators from that Salvador NGO practiced a form of critical education that was not primarily aimed at changing the children but was based on observing, listening, spending extended periods of time with the kids, and respecting and valuing them for their human assets, their remarkable skills, and their culture of resistance. This, as Veissière saw it, was a revolutionary practice that had courageously steered away from quick fixes and telegenic street cleanups, and was instead dedicated to slow and in-depth systemic change based on historical and political conscientization "from below." It was a revolutionary practice based on a qualitative process that could only be called ethnography.

Conclusion: Toward a Participatory Ethnographic Praxis

So far, we have argued that most of the popular, nongovernmental, and nonformal institutions that have been put in place to "rescue" or empower street children fail in their mandate at best, often simply reinsert nurturing children into the silent status quo of wretched poverty, and at worst reinforce and accentuate the pain and experience of marginalization of children who already suffer from social exclusion. We have also seen that these conclusions were most often reached not by NGO personnel, but by ethnographers, who by definition are positioned ambiguously, but also more fluidly, as insiders and outsiders in the various social, political, interpersonal, cultural, and institutional sites that come into play in this grim picture. Typically, ethnographers benefit from hindsight, a systemic view, trust, and a plurality of microperspectives and narratives of lived experience that are hidden to the vast majority of local actors, who are confined to their habitus and to mutually exclusive specific social and spatial positions (Markowitz, 2001). Thus, we argue that these multiple perspectives inherent to an ethnographic paradigm are lacking from most popular education initiatives and structures, and that they need to be more actively incorporated in such programs.

An ethnographic paradigm, thus, could revitalize popular educational institutions and programs on two counts: critical/theoretical, and pedagogical/practical.

As already mentioned, adding an ethnographic dimension to popular education places the children's lived experience at the center of the intervention. Such a dimension seeks to not only critically understand the realities and constraints faced by these children but also situate them within broader historical, social, cultural, and economic structures, thus trying to understand as well as address the systemic causes of the "problem" instead of focusing on the symptoms. In the Brazilian context, it is thus imperative to understand and tackle the cultural and economic issues (the iceberg) that produce street children (the tip of the iceberg) and to situate these issues within a global neocolonial framework (the network of seas and weather that produce the icebergs).

This conscientization to local and global political mechanisms should then be followed by action at micro and macro levels. In the introduction to this chapter, we showed how the polysemic nature of ethnographic thinking prompted us to identify "children," and particularly "street children," as resilient cultural producers and "whole beings" who were not only capable but also in need of drawing upon their own modes of knowledge and visions to mediate social change through popular education. Popular education's focus on organizing people to take collective action (Kapoor, 2004), therefore, should not be forgotten. Thus, the task of the ethnographer-educator is to draw upon insights such as those enunciated in this chapter, and to contribute to the creation of spaces where the disenfranchised can articulate their own visions of social change and contribute to more inclusive transformations of civil society. This might entail revitalizing and extending the local ramifications of the National Movement of Street Boys and Girls (Movimento Nacional de Meninos e Meninas de Rua, MNMMR), whose 1989 Second National Congress in Brasilia to denounce poverty and police violence had led to the constitutional formulation of a Children and Adolescent Act (Hecht, 1998). The MNMMR was founded on the ideas of conscientization and citizenship. Its central goal was to invite street children to rechannel the courage and energy spent in surviving in the streets into courage and energy spent in organizing themselves to challenge the oppressive superstructure that allows an otherwise warm society to go on carnivaling without creating a safety net for disenfranchised children. Whereas the MNMMR made successful strides toward individual civil rights for disenfranchised children (Children and Adolescent Act), it was not able to engage the sociological imagination in embracing civil responsibility for children without homes. After 500 years of history, Brazil still does not have a state-sponsored system to care for orphans and runaway children. Disenfranchised youth are at the mercy of religious and charity groups, with their patronizing and theological twists. It is here, in our view, that critical ethnography, popular education, and the social predicament of street children intersect.

On a concluding note, we would like to stress once more that we are advocating a multidisciplinary praxis between ethnography and popular education, both of which we see as indispensable components for any process of emancipatory social change.

As we argued throughout this chapter, most nonformal/popular educational and social institutions that seek to assist street children lack the critical perspective to comprehend and address the complex issues in which the lives of these children take place. While ethnographic research can be invaluable in shedding light on the causes and conditions of poverty and the failure of social institutions, we should also note that it has almost always failed to produce positive change for those who experience these predicaments. Critics have often called attention to the fact that ethnographic investigations usually remain within the realm of the descriptive, never travel beyond elitist academic circles, and produce no tangible change in the lives of the marginalized individuals whose lives are being "theorized," while scholars gain academic prestige through the misery of "their" research subjects (Fine, 1994; Smith, 1999). Commenting on his own fieldwork, Hecht (1998) remarked that "suffering is the stuff that keeps many foreign visitors employed" (p. 17), and pointed to a phenomenon he labeled "street-children tourism," which feeds on an "industry of researchers that treat street children as raw material there for the taking, [and translate] the anguish of real children into [documents] that benefit above all [themselves]" (pp. 18–19).

In light of the risks inherent in both extremes of pragmatic sightlessness and theoretical aloofness (and at a risk of grossly simplifying this picture), we call upon critical educators and ethnographers who are concerned with issues of social change to share and unite their visions and practices in order to generate a collaborative ethnographic praxis of popular education.

References

Arnove, R., Franz, S., Mollis, M., & Torres, C.A. (2003). Education in Latin America: Dependency, underdevelopment, and inequality. In. R.F. Arnove & C.A. Torres (Eds.), *Comparative education: The dialectic of the global and the local* (pp. 313–337). Lanham, MD: Rowman & Littlefield.

Childress, H. (2004). Teenagers, territory, and the appropriation of space. *Childhood, 11*(2), 195–205.

Cubillos, R.H. (2002). *Braving the streets of Brazil: Children, their rights, and the roles of local NGOs in northeast Brazil.* Unpublished doctoral dissertation, University of Southern California.

Diversi, M. (1998). *Street kids in search of humanization: Expanding dominant narratives through critical ethnography and stories of lived experience.* Unpublished doctoral dissertation, University of Illinois, Urbana-Champaign.

Diversi, M. (2003). Glimpses of street-children through short stories. In. K.J. Gergen & M.M. Gergen (Eds.), *Social construction: A reader* (pp. 109–119). Thousand Oaks, CA: Sage.

Fine, M. (1994). Working the hyphens: Reinventing self and other in qualitative research. In N.K. Denzin & Y.S. Lincoln (Eds.), *Handbook of qualitative research* (pp. 70–82). Thousand Oaks, CA: Sage.

Freire, P. (2000). *Pedagogy of the oppressed.* New York: Continuum.

Goffman, E. (1961). *Asylums: Essays on the social situation of mental patients and other inmates.* Garden City, NY: Anchor Books.

Goldstein, D.M. (2003). *Laughter out of place: Race, class, violence, and sexuality in a Rio shantytown* (No. 9). Berkeley: University of California Press.

Hanchard, M. (1999). *Racial politics in contemporary Brazil.* Durham, NC: Duke University Press.

Hecht, T. (1998). *At home in the street: Street children of Northeast Brazil.* New York: Cambridge University Press.

Kapoor, D. (2004). Popular education and social movements in India: State responses to constructive resistance for social justice. *Convergence, 37*(2), 55–64.

Markowitz, L. (2001). Finding the field: Notes on the ethnography of NGOs. *Human Organization, 60*(1), 40–46.

Mickelson, A. (Ed.). (2000). *Children on the streets of the Americas: Globalization, homelessness and education in the United States, Brazil, and Cuba.* New York: Routledge.

Scheper-Hughes, N., & Hoffman, D. (1998). Brazilian apartheid: Street kids and the struggle for urban space. In N. Scheper-Hughes & C.F. Sargent (Eds.), *Small wars: The cultural politics of childhood* (pp. 352–388). Berkeley: University of California Press.

Smith, L.T. (1999). *Decolonizing methodologies: Research and indigenous peoples.* London: Zed Books.

Veissière, S. (2006). Tropicalism: Misplaced logocentrism and the production of tropical(ist) identities in postcolonial Brazil. In K. Pramela et al. (Eds.), *Discourses on culture and identity: An interdisciplinary perspective.* Kuala Lumpur: UKM Press.

Veissière, S. (2007). *Hookers, hustlers, and gringos in global Brazil: The transnational political economy and cultural politics of violence, desire, and suffering in the streets of Salvador da Bahia.* Unpublished doctoral dissertation, McGill University Montreal, Canada.

CHAPTER 13

Citizens Educating Themselves: The Case of Argentina in the Post–Economic Collapse Era

Luis-Alberto D'Elia

> A review of the literature reaffirmed that research and visions related to Adult Basic Learning and Education in the South are dominated by the North, by international agencies and by English-speaking reviewers, often ignoring or dismissing research produced in the South, especially if it is written in languages other than English.
>
> Torres, 2004

In reviewing some of the particular contexts in which hundreds of thousands of Argentineans have been organizing and educating themselves around and after the time of the country's recent historical crisis, I will be looking at relevant work done by researchers and educators, paying special attention to those coming from the South, in particular from Argentina. The intention of this chapter is to look critically into the exceptional socio-cultural-political conditions that enable Argentinean "crisis" new movement groups to seek out and practice uncompromising, autonomous ways of adult and collective nonformal and informal education.

Introduction: No Confidence in the Old System

Argentina, a country that after World War II was considered the "grain supplier of the world" and that at one time used to have one of the most stable social systems (under the Peronist era), collapsed socio-politico-economically and very dramatically at the beginning of the third millennium.

As the crisis almost completely wiped out the country's middle-class majority out of the social spectrum and set a historical precedent, Argentina defaulted in the payment of its foreign debt to the International Monetary Fund (IMF) and attached organizations. In fact, in December of 2001, the government defaulted on about $140 billion of debt, the largest sovereign debt default in history (Feldstein, 2002). Following that, Argentina's currency (the Peso) and its banking system collapsed, and the country sank further into depression. Preceding the collapse, the shaky Argentinean government had sequestered the savings of the once-dominant and strong middle class, immediately sending this large group of now-destitute people to join the already growing number of poor and under classes. Clearly, the country was in a complex and multifaceted crisis (Armelino et al., 2002; Klein, 2003; Lodola, 2003; Monteagudo, 2004; Palomino, 2004; Weisbrot, 2005;). However, the collective response to the crisis began simmering and saw the quick formation of an unprecedented social movement that not only managed to depose five country presidents in a single week but also invented a new form of socio-economic-organizational survival scheme that surprised the Argentinean crisis analysts around the world (Klein, 2003; Lodola, 2003;). This chapter speaks about the characteristics of this movement and provides some insights into the unique historical forms of citizens' informal and autonomous education.

The Social Groupings of the Crisis

As Klein (2003) has noted,

> With all of their institutions in crisis, hundreds of thousands of Argentinians [sic] went back to democracy's first principles: neighbors met on street corners and formed hundreds of popular assemblies. They created trading clubs, health clinics and community kitchens. Close to 200 abandoned factories were taken over by their workers and run as democratic cooperatives. Everywhere you looked, people were voting. (p. 1)

The crisis generated a popular movement whose important characteristic was a purposeful effort to be autonomous and independent of any organized sociopolitical structure. Whether they were *piqueteros* (that is, groups of unemployed, economically disenfranchised middle class), food rioting groupings, or neighborhood assemblies, these Argentinean associations were radically opposing past and current sociopolitical experiments (Armelino et al., 2002; Klein, 2003; Lodola, 2003). More importantly, they were securing an unprecedented change in the way social movements would relate to social agency, in general, and to the established powers in the world, in particular. They resisted, with huge, visible, and tangible successes, government attempts to assimilate them into its structure, and were not willing to join any specific political party or other established social and labor movement, including nongovernmental organizations (NGOs). The new social arrangements clearly had a communal consciousness of fighting the corrupt government and planning for life in self-sustainable and self-sufficient ways that were ready to operate by dispensing with the country's political leaders who were incapable of doing anything and therefore, in the people's minds, invisible. At the frequent massive demonstrations during the deep crisis, for example, hundreds of thousands of Argentineans would chant, "*¡que se vayan todos!*" ("all of them [all state representatives, politicians] leave!"[1]) (Klein, 2003, p. 1, see also Armelino et al., 2002; Auyero & Moran, 2001; Palomino, 2004). This extraordinary rebellion against assimilation of and dependency on old structures appears to mark a parallel autonomy by new movement members to teach and learn by themselves.

Autonomous Groups, Creative Pedagogy

In looking back at these situations and even observing current social structures, one can see that there are hundreds of autonomous organizations in Argentina. Their autonomy from institutional and political powers is unique and was virtually nonexistent in sociopolitical groups in Argentina during the 1980s and the 1990s. Monteagudo (2004) predicts that these groups will likely stay this time, transcending the temporality that characterized similar ones in previous crises. And although there were no organized or even street-based educational projects that might have "conscientized" the rebelling public this time, clearly there were spontaneous and informally located learning interactions that people have quickly shared and immediately operationalized to achieve a viable social movement vis-à-vis the powers-that-be. But again, this type of informal education might not last if it is not sustained via an organized social program that also gives important and long-term meaning to both the formations of the new social movements and the historical affirmation of those movements. As Monteagudo notes, the task still needs more work, for, thus far, "they have not been able to stabilize that energy in structures that will enable them to grow and succeed. Perhaps, if these autonomous groups continue to prosper in the future and are able to unite in decentralized structures, that will change" (p. 1). With autonomous, sustainable social grouping, we can hope for a pedagogy that is characterized by its independence from institutional, political, and structural powers and interests. As noted above, some of that pedagogy has at least started to become more evident in the recent years.

Sterilization of Reproductive Education: Discontinuation of Capitalism Profiting from Education

According to Weber, power is the realization of human will even against the resistance of others (as cited in Murphy, 1988). We could say more about the sociological concept of power by referring to Baldus (cited in Murphy, 1988) definition, where power can be seen as the ability of a "center unit," in a structure of social inequality, "to maintain, reproduce, or reinforce" its position with respect to a "periphery unit" (p. 188). Further, Murphy defines a very subtle power subcategory: the power of a unit "to profit from," which is defined as the "capacity to take advantage of possibilities that are presented by others" (p. 148). In the particular case of society-school relationships, Murphy explains that the bourgeois class has "the *power to profit* from educational knowledge and to constrain . . . the definition of what counts as educational knowledge" (p. 148) (my italics). To what degree is this type of almost imperceptible power influencing educational process in the new Argentinean movements?

Further, considering that most sociological theories of education assume that the school and educational systems are highly dependent on wider society (that is, society's dominant groups) and that the latter has demonstrated power over schooling (Parsons; Bowles, Gintis, & Young, as cited in Murphy, 1988), it is important to examine how the Argentinean autonomous movement enters into the education-society power equation. When attempting to draw any conclusion from these power relationships, one has to consider that Argentinean society has not been an exception to the hegemonic capitalist and neoliberal model, and that it is still debatable whether the dominant groups in society have the power to profit from education *because* the educational (schooling) systems have adapted to the capitalist system (Murphy). Regardless of the validity of the latter point, in the current autonomous movements in Argentina, the opportunities for the dominant groups in Argentinean society to have the power to profit from the education for their members will be directly proportional to the degree of assimilation of the new groups to the institutional, labor, and political structures of the country.

Another opportunity for Argentinean capitalists to profit from their groups' efforts to educate and train their members is by incorporating these members in their production machinery (employing them under the regular market rules). However, for the autonomous groups that emerged around and after the crisis, a variety of situations exempt them from having to negotiate their education and their organizing efforts with the established capitalist system. These situations would include the fact that a good number of the assembly members (*asambleistas*) are not entering the capitalist labor force but are creating their own microenterprises (home business) (Klein, 2003; Lewis, 2004). In addition, some of the groups of piqueteros have decided to continue working on the state-plan program, escaping, in this way, dependency on the private capitalist employers (Lodola, 2003). And finally, the unemployed masses who took over the control of the abandoned factories are running them as cooperatives and are rejecting organizing their businesses to a profit distribution system similar to the capitalistic approach (Klein).

Consequently, it appears that the characteristic of the autonomy of the piqueteros, asambleistas, and hundreds of thousands of Argentineans organized around and after the crisis protects them from being vulnerable to the society-education power relationship. Specifically, Argentinean workers who escaped being regulated by the capitalist economic interests are *agua para su propio molino* (bringing water to their own well) and are serving their own educational needs.

Liberating from State Dependency

Klein (2003) argues that the popular neighborhood assemblies that congregated as a large number of laborers, unemployed workers, and middle class kept their commitment to be autonomous from any previously organized, and many times corrupt, structure. Participants of these assemblies would negotiate all sociopolitical terms with much care in protecting their independency. Klein was a witness to the outstanding autonomy of these groups. It is clear that any invitation by politicians to discuss social issues would be rejected, except when the politician would come to where the assemblies were congregated, that is, the streets or other public places at the nationhood corner. The assembly members did not allow any indirect representation of their popular demands because they suspected that middle persons may be, one way or another, attached to politicians, brokers, or gatekeepers, or even one of their own could water down their demands or even betray them, a reality they know too well that happened many times before the crisis. Undoubtedly, the potential of these autonomous groupings for creating new educational ways, independent of institutionalized pressures, is exceptional.

Independency in a Traditional State-Patronizing Nation

One can argue that the social groups' dependency on the state has been a characteristic of the Argentinean society since the Peronist regime, and that dependency will likely characterize the new groupings, in spite of the crisis. Even though a fundamental sociopolitical arrangement in the form of the cliental system led by *punteros* (social brokers) was present in Argentina during the crisis, which probably has been inherited from the Peronist regime, at the time of the crisis the state failed to sustain it (Auyero et al., 2001). The punteros, who traditionally mobilized the poor and working-class neighborhoods for elections and other political activities and who were in charge of the distribution of state welfare benefits and, likely, educational strategies at the neighborhood/community level, were still playing their role during the crisis. However, at this time, their political/institutional relationship changed. One illustration comes from the punteros' activities during the 2001 food rioting. During the violent weeks of the riots, they looked for food outside the usual places and did not use the usual means. In fact, the traditional state programs that had been the sources of food for the punteros' clients were not responding, and the brokers turned their attention to local supermarkets.

Certainly, as the crisis continuously intensified in Argentina, the majority of the people were left with nothing to lose and possibly something to gain. They started to restore their sense of autonomy and to remove their support base from the deceiving structural powers and transfer it to their own communities. Many Argentineans chose their own path to create new ways of organizing, teaching, and learning.

Outside the Academia

Most of the piqueteros, asambleistas, factory-take-over workers, and other Argentineans congregated around hundreds of new sociopolitical groupings and undertook teaching/learning with a similar philosophical approach used in their organization: autonomous, self-reliant, independent of external interest and powers, and strictly locally and regionally driven and mobilized.

If the educational experiences of these Argentinean crisis groups are characterized by autonomy and independence from established institutional and political interests and powers, it is evident that these groups will not use formal systems of education but rather informal and nonformal ones. At the same time, it is also understandable that in times of deep institutional crisis, when basic socio-economical-political tenets are questioned, as they have been in the case of Argentina, the most nonconventional extrainstitutional forms of education will be adopted by group members. Consequently, we may expect the informal education category to be the most popular amongst group members. In defining informal learning, Livingstone (cited in Schugurensky, 2000) writes that it can be seen as "any activity involving the pursuit of understanding, knowledge or skill which occurs outside the curricula of educational institutions, or the courses or workshops offered by educational social [or corporate] agencies" (p. 1).

Informal Education

Among others, piqueteros and the assembly movements have been exploring strategies of informal education. However, the important characteristics of their educational approach have been their uncompromising autonomy from established forms of organized education. One important reason for the new movement to take such an anarchist positioning has been explained before: their absolute rejection of the organized governmental or nongovernmental structures during the crisis in Argentina.

How has informal education been implemented by the participants of those movements? Besides the highly valuable informal learning that happens in the lives of citizens as they organize themselves to attend to their basic survival needs, members of piqueteros or assembly organizations organized their education with more intentionality and started asking the popular educators to organize immediate and needed areas of learning. One such educator's group organized the Área de Educación Popular del Movimiento Barrios de Pie (Movement Barrios de Pie's-Neightborhood Standing Up's-Popular Education) in 2002. This focused on collective projects such as "popular education with children, literacy and post-literacy; elementary and high school completion" (p. 1); workshops (in Argentinean and Latin American history and political education) for leaders and for participants of neighborhood soup kitchens; a diversity of other workshops, such as trade work, popular assemblies' participatory techniques; and traveling workshop discussions on the proposed Free Trade of the Americas (FTA) (ALCA in Spanish) and foreign debt, amongst others (Movimiento Barrios de Pie, 2005). In fact, sociopolitical themes that are part of the context of their existence and survival were never absent; moreover, as evidenced later, these are generic parts of the people's daily existence.

Nonformal Education and Literacy Programs

Literacy programs in Latin America in the 1970s were characterized by a critical approach: raising the consciousness of illiterate adults and educating the participants to question predominant social and political oppressive structures. Ligas Agrarias, which especially educated the farmers in understanding, among other things, the underlying factors of their poverty and dependency, was one such program. Programs like these usually are part of nonformal education, due to their structure and planning and the participation of formal social and development agencies in their implementation. However, there is a need to examine the influence of the particular circumstances of the Argentinean crisis on the literacy programs in order to identify possible differences between the literacy programs during the crisis and more traditional programs.

Some of the literacy programs implemented in Argentina in the recent years use the Freirean philosophical approach as evidenced in the use of *popular education* pedagogy. In a popular education process, learners learn about their own sociopolitical oppressive situation using their own expressions and are encouraged collectively to become active participants in the process of their own liberation (Freire, 1970). Freire's popular education is critical education with its assumption of the crucial role that education plays in uncovering and changing society's unjust structures of oppression.[2]

In a recent national meeting of popular educators in Argentina, it became clear that the popular education programs targeted illiterate adults, and other learning groups covered a number of group-relevant issues and strategies. At the same time, there was a general conceptualization of popular education as "popular" not only in terms of free access, with a bottom-up approach that is community oriented and non-governmental in nature, but also in terms of its informality. As Coco, one popular educator from Buenos Aires, put it,

> The popular educator is not a person that learns about how to educate from books. Popular education is not a thing that it can be found [only] in books, it is a concept. Popular education is a political act: it is present every day wherever there are *compañeros* [comrades] in the struggle. Popular education is there. There are also the popular educators: They are in the *piquete* [picketing], in the marches, in the students center. Popular

education is in every corner where we meet and where we take decisions. (Movimiento Barrios de Pie, 2002)

A more traditional (massive) literacy program recently borrowed from Cuba by two Argentinean popular education organizations is the *"Yo, sí puedo"* ("I am surely capable"). The program caught the attention of the organizers after visiting the program overseas and upon realizing the huge task ahead of implementing popular education in Argentina. It is a literacy-based audiovisual program for adults provided to the Argentinean Barrios de Pie by IPLAC (Instituto de Pedagogos de Latinoamérica y el Caribe) and by the UMMEP (Un Mundo Mejor es Posible) (Movimiento Barrios de Pie, 2002, 2005). Will this more structured and prepacked program, structurally organized educational strategy, work for the vast majority of those who have lost trust in any structured or institutionalized help?

Education and New Trade Unions

In speaking about structured formal education vis-à-vis the needs of workers, Livingstone and Sawchuck (2005) wrote that

> working-class peoples' indigenous learning capacities have been denied, suppressed, degraded or diverted within most capitalist schooling, adult education institutions and employer-sponsored training programs, at the same time as working class [sic] informal learning and tacit knowledge are heavily relied on to actually run paid workplaces. (p. 110)

Even though this research has been done in North America, a possible generalization of the work can be applied to the Argentinean workers since the Argentinean government (most importantly, under the Menem administration in the late 1990s) structured the functionalities of the country in line with the so-called knowledge-based economy. In this economy, *expert* knowledge is validated over workers' indigenous learning. Under this assumption, most workers would be considered to have educational *deficits* that, according to Livingstone and Sawchuck, reflect the human capital concept of education relevant to the knowledge-based economy, even if trade unions, nevertheless, have been the "primary sites of working class [genuine] learning" (pp. 110–111).

The Argentinean trade unions have been historically organized around a central national confederation, the General Confederation of the Workers (CGT, Central de Trabajadores Argentinos). This traditional

(Peronist) organization has been criticized as corrupt and siding with the country's economic and political powers. Other confederations and labor unions have emerged in the recent history of the country. One of them, popular around the time of the recent Argentinean crisis, is the Central de los Trabajadores Argentinos (CTA), an alternative national union confederation to the CGT. The CTA defines itself as a "new confederation of workers, employed or unemployed, founded on three essential concepts: direct affiliation, outmost democracy, and political autonomy" (CTA, 2002).

We can expect informal and nonformal learning to happen through and as a result of the trade unions. However, around the time of the crisis, the level of confidence in organized labor might have decreased significantly for many impoverished Argentinean unionized workers.

Indeed, even today, structured and institutionalized assistance would be rejected by Argentineans as they have experienced a long history of deception by the government and its institutions. However, if pavement help is perceived as an extension of their labor struggles through noncorrupted unions, the Argentinean workers may trust this helping hand. This may be the case with the CTA. Its educational programs are organized under the Instituto de Estudios y Formación (Institute for Studies and Training). As things are now, this could constitute one of the institutional nonformal projects of education that is acceptable to some Argentineans in the crisis movement.

Another popular and quasi-institutionalized nonformal education project is the one offered through the Roman Catholic Church's predominant social agencies such as Cáritas (Cáritas, 2005). Throughout its different programs and outlets across the country, such as "the community services centers, the lunch programs/kitchen soups, the parishes and chapels," there are nonformal education initiatives and activities servicing communities, families, and individuals of all ages and from diverse socioeconomic backgrounds (Cáritas). The pedagogical objective is to encourage people's capacity to be "critical and *autonomous*" (my italics) (Cáritas). In the case of children, Cáritas programs tend to socialize them into the mainstream schooling in Argentina. Nevertheless, in this case, the church's agency understands the need to go beyond the formal education offered by schools. In spite of the highly structured institution to which it belongs, some of the Cáritas chapters (in some provinces) have been able to selectively respond (quite faithfully) to the needs of the poor and the negatively affected middle class. As such, they have developed a good reputation, not only among the communities they serve but also among the general population, especially in some regions of the country. Cáritas has been critiqued for being *asistencialista* (providing only band-aid assistance and not dealing with the structural roots of the social problems) by other social agencies that might be suspicious of any pronounced disengagement of Cáritas from the powers-that-be during the crisis. To its credit, though, Cáritas was one of the few mainstream institutions trusted by the disillusioned Argentineans in some areas of the country. This may be due, in part, to the high level of community-based development and participation that Caritas has in its programs.[3] Indeed, with the amount of experience in living through both locally and globally induced oppression, complemented by the informal training that workers and progressive trade unions have received to fight against the central government and its formalized institutions, citizens' movement should be able to see and recognize any good work that groups such as Cáritas are undertaking with genuine interests in the lives of Argentina's disenfranchised.

Conclusion

In this chapter, my intention has mainly been to attempt to present some unique educational analyses pertaining to the situation of Argentina especially since the economic crisis during the early years of the third millennium. A deliberate perspective here has been to speak with a voice that should situationally represent the needs and the feeling of the public in overcoming the difficult situation in which they found themselves after the sudden loss of purchasing capacity, indeed, economic viability, following this globally induced crisis. Specifically, this chapter tries to look critically into the exceptional socio-cultural-political conditions that enable Argentinean new movement groups that emerged during and after the crisis to search for and practice uncompromising, autonomous ways of nonformal and informal education. The reality is that, in the lives of the common man and woman, the recent Argentinean crisis had apocalyptical proportions and generated a collective response that unleashed an unprecedented social movement that resisted assimilation by government, political parties, and established social and labor movements, even NGOs. This resistance and autonomy is accompanied by a parallel autonomy by the new movement members to teach and educate themselves. With that newfound autonomy, the new movement not only broke off from the deceiving powers-that-be of the past, creating new organizational and educational spaces, but it also managed to escape the Argentinean capitalist's power to profit from the education of the new movement's members. But what forms have this educational process taken?

On the one hand, piqueteros, the assembly, and other crisis movements, amongst others, have been exploring strategies of informal education. However, this informal education takes a particular shape in these popular movements. Through their "popular educators," piqueteros or assembly organizations have been focusing not only on children's popular education, literacy, and postliteracy, but also on history and political education for leaders and for participants of neighborhood soup kitchens, trade work, and popular assemblies' participatory techniques. Lately, these educators have organized discussions on the proposed FTA, foreign debt, and other topics. In fact, sociopolitical themes are an integral part of these historic social movement shifts and should affirm their survival and long-term existence. Moreover, it appears that the critical examination of the sociopolitical context in which these movements are immersed is a prerequisite for these groups' educational undertaking. The nonformal educational strategies, on the other hand, have been used for literacy programs that traditionally in Argentina (as well as in other Latin American countries) have used popular education that follows the Freirean educational platform. Following the crisis, a particular literacy program was implemented by Barrios de Pie, a crisis group, in collaboration with Cuban and Caribbean educational organizations. Other nonformal education strategies come from an alternative trade union confederation. Even though Argentineans organized in postcrisis movements reject any structured institutionalized assistance, if the help is perceived as part of their labor struggles through their noncorrupted unions, it may be accepted. Moreover, the national workers confederation, CTA, has been providing nonformal education that, far from creating the usual labor dependency, is encouraging its members to gain consciousness about their labor situation.

Looking Forward

In sum, it is now clear that there are hundreds of organizations in Argentina that are basically autonomous from institutional and political powers (Lodola, 2003; Monteagudo, 2004) and whose informal and nonformal education programs go beyond learning how to organize and survive in the market economy. Indeed, their members are using popular education to understand their own sociopolitical context to counterweigh and resist assimilation by the established political system that failed them in the process leading to the crisis. Their long-term survival may be more likely than in previous crises, and this sustainability will probably ensure a more effective and systematic informal educational process. This educational process will, in turn, become more important as other countercurrent (antiglobalization) movements around the globe look at the Argentinean autonomous groups and learn more about how they can create and manage progressive educational possibilities without compromising their autonomy that, in the Argentinean case, guarantees independence from corrupt centralized sociopolitical arrangements. In a recent visit to Argentina, Michael Hardt restated that we are living in the final times of imperialism (although a qualified one), and that what makes imperial hegemony vulnerable is not anti-imperialism but the self-management by autonomous groups like the ones in Argentina, Mexico (Chiapas), and Brazil (landless movement).

Notes

1. Translations from the originals in Spanish have been done by the author.
2. My own personal experiences with Freirean pedagogy in 1970 at the Federal University of Santa Maria, Rio Grande do Sul, Brazil, would confirm this.
3. My personal experience in working with Cáritas in Entre Ríos Province for many years. In the early years of the third millennium, I have met with some community program participants, natural leaders, and Cáritas management, and I have visited specific community projects with an educational component in Paraná City and in small towns and villages of Entre Ríos Province.

References

Armelino, M., Bruno, M., Larrondo, M., Patrici, N., Pereyra, S., Perez, G., et al. (2002). *La trama de la crisis: Modos y formas de protesta social a partir de los acontecimientos de diciembre de 2001.* Facultad de Ciencias Sociales, Universidad de Buenos Aires (Informes de Coyuntura, N° 3). Retrieved October 15, 2005, from http://www.iigg.fsoc.uba.ar/docs/ic/ic3.pdf

Auyero, J., & Moran, T.P. (2001). The dynamics of collective violence: Dissecting food riots in contemporary Argentina.

Cáritas Argentina. (2005). *Educacion.* Retrieved November 15, 2005, from http://www.caritas.org.ar/htm/educacion01.htm

Central de Trabajadores Argentinos. (2002) Retrieved December 2, 2005, from CTA Official Web page: http://www.cta.org.ar/institucional/institucional.shtml

Feldstein, M. (2002, May 28). Argentina doesn't need the IMF. *The Wall Street Journal.*

Freire, P. (1970). *Pedagogía del oprimido.* Mexico: Siglo Veintiuno.

Klein, N. (2003, May 12). From pots to politics. *The Guardian.* Retrieved October 22, 2006, from http://www.guardian.co.uk/Columnists/Column/0,5673,953825,00.html

Lewis, A. (Director). (2004). *The Take* [film]. Argentina Barna-Alper.
Livingstone, D., & Sawchuck, P. (2005). Hidden knowledge: Working class capacity in the "knowledge-based economy." *Studies in the Education of Adults, 37,* (2), 110–122.
Lodola, G. (2003). *Protesta popular y redes clientelares en Argentina: El reparto federal del Plan Trabajar (1996–2001).* Pittsburgh, PA: University of Pittsburgh, Departamento de Ciencia Política.
Monteagudo, G. (2004, July 25). The Argentinean social movements and Kirchner. *Blue III,* 13. Feature Archive *BlueGreenEarth.* Retrieved November 27, 2005, from http://www.bluegreenearth.us/archive/article/2004/monteagudo-1-2004.html
Movimiento Barrios de Pie. (2002). *¿Qué es el Área de Educación Popular?* Retrieved November 10, 2005, from http://www.barriosdepie.org.ar/
Movimiento Barrios de Pie. (2005). *Encuentro Nacional de Educadores Populares.* Retrieved November 10, 2005, from http://barriosdepie.bitacoras.com/archivos/2005/05/25/primer-encuentronacional-de educadores-populares-del-movimiento-barrios-de-pie
Murphy, R. (1988). *Social closure: The Theory of monopolization and exclusion.* Oxford: Oxford University Press.
Palomino, H. (2004). *Un análisis de la "economía moral" del movimiento autogestionario* (Revista Herramienta La Fogata). Retrieved November 15, 2005, from http://www.neticoop.org.uy/documentos/dc0396.html
Schugurensky, D. (2000, October). *The forms of informal learning: Towards a conceptualization of the field.* NALL (New Approaches to Lifelong Learning) Working Paper #19. Retrieved December 2, 2005, from http://fcis.oise.utoronto.ca/~daniel_schugurensky/
Torres, R.M. (2004). Lifelong learning: A new momentum and a new opportunity for adult basic learning and education (ABLE) in the South. *Convergence, 37,* 3.
Weisbrot, M. (2005, October 12). Good news at last! The IMF has lost its influence. *Counterpunch.* Retrieved October 27, 2005, from http://www.counterpunch.org/weisbrot10122005.html

CHAPTER 14

A Freirean Analysis of the Process of Conscientization in the Argentinean Madres Movement

Charlotte Baltodano

In 1976 the Argentinean military junta staged a violent coup that removed the elected government and established a regime that relied on large-scale human rights abuses to squelch resistance. Legalized violations of human rights were enacted by paramilitary groups, civil forces called *milicos,* resulting in a policy of systemic kidnappings and assassinations of "suspicious" targets. The term *desaparecidos* (the disappeared) came to be used to identify the thousands of workers, students, and leftist activists who, once abducted, never returned to their homes. They were taken from their homes, illegally detained, and sent to clandestine incarceration centers, where they were tortured, murdered, and their bodies secretly disposed of (Amnesty International, 1981). Human rights groups estimate the number of the disappeared to have reached 30,000. The indiscriminate nature of the kidnappings and the impunity with which they were carried out brought forth terror, gripping the country with fear and deafening silence. The rule of law no longer applied.

The military junta repressed all possibilities of resistance and communal organization through brutal force, justifying the violence with the need to enforce "order." Armed paramilitary forces assisted the regime in maintaining a state of siege. With acts of indiscriminate repression, they paralyzed civil attempts to respond to the injustices, leaving the population incapable of resisting. However, despite the overwhelming oppression, voices of resistance against the military regime eventually burst out with great courage. These were the voices of the mothers of the desaparecidos, who demanded information about the whereabouts of their missing children. The outspoken challenge of the Madres de Plaza de Mayo (Mothers of May Square),[1] as their movement came to be known, became a symbol of resistance against human rights abuses. Hebe de Bonafini, president of the Madres movement, recounts the beginning of their resistance in 1977 at the Plaza de Mayo:

> We would wave our scarves and shout to the *"milicos"* that we had "disappeared" children. There was no other thing we could do, only inconvenience . . . the military in this way. Videla, then dictator, sent an emissary (that afternoon) to notify us that if we leave the Plaza, he would agree to talk to us . . . we stayed and stood side by side, holding each other, holding each other in a column. (Later, that day) . . . Videla sent *"milicos"* of war armed with helmets who ordered us to vacate the Plaza. We would not leave. Then one of them ordered to point their rifles at us and when they yelled *"Apunten"* (Aim), we shouted *"Fuego!"* (Fire!). (de Bonafini, 1988)

Their determined voices of defiance and courage resonated around the world. They were able to organize themselves, mobilize organizations, and maintain a sustained movement for human rights. The Madres movement began with individual mothers whose main concern was to get information about their children. Slowly, they became aware that, in order to receive information from the military, they needed to organize themselves. As a collective they created new forms of actions that opposed and challenged the military's abusive practices. Beginning with fourteen members, the collective grew to comprise 5,000 (Guest, 1990). The mothers had succeeded in creating a national network that evolved into an alternative for justice. Their activities grew to encompass assemblies for the resolution of psychological problems (with the help of a team of psychologists), participation in anti-imperialist tribunals and marches for the defense of life, and even helping peasants win land disputes. Later, their presence was secured in judicial bodies such as the military tribunals in Argentina, as well as the criminal trials against Astiz in France.[2]

However, this activism meant that, through the years many mothers themselves disappeared, others were kidnapped and beaten, and many had their valuables broken and stolen during raids of their homes. Despite these acts of intimidation, the Madres, along with the Grandmothers of the Plaza de Mayo, the Movement of Relatives of the Disappeared, and other human rights organizations, persisted in their denouncement of repression and terror, joined by solidarity groups around the world. On a global scale, the Madres movement became a symbol of the denouncement of authoritarianism and reclaiming of democracy. Their collective practices have influenced solidarity groups in France, Holland, and Spain, to name a few, seeking to this day alternatives to uncovering oppression.

The Madres represent one of the largest and most sustained civil movements in Latin America. They have attracted researchers, educators, and scholars, who have studied the Madres from different perspectives. From the perspectives of human rights and NGOs (Clark, 2001; Keck & Sikkink, 1998; Sikkink, 1993), pedagogy, democracy and feminism (Hernandez, 1997), feminist social movement (Calderon, Pisscitelli, & Reyna, 1992), space and symbol, identity and social movements (Feijoo & Gogna, 1990; Hernandez, 1997; Jelin, 1985; Jelin, 1990), biographical accounts (Bousquet, 1983; Guzmán Bouvard, 1994), political subjects, and ultimately into the symbol of resistance to military dictatorship (Navarro, 2001), to name a few. Their multifaceted analyses have helped generate insights of the experiences of the Madres as mothers/women/symbols/private and public space/social movement/identity/memory/culture.

These researchers have used various foci to enrich the understanding of the Madres' role in establishing identities and transmitting ideologies, retracing their social positioning and their development as social actors, and, of course, dealing with their bravery and audacity. They have also analyzed the movement as a new ethical response (Calderon, Pisscitelli, & Reyna, 1992; Feijoo & Gogna, 1990) to human rights abuses. Yet, they have given little serious attention to understanding the learning dimensions of the Madres and the role they played in creating new responses to situations within their social movement. Thus, I focus on the political consciousness changes that have resulted from the actions of the Madres to denounce the human rights abuses and continue their work against multiple conditions of oppression to this day. Griff Foley's (1999) wise statement comes to mind: "For people to become actively involved in social movements something had to happen to their consciousness—they must see that action is necessary and possible" (p. 103). In the Madres movement there is a complex dynamic of learning that goes from being a personal and individual struggle to becoming a collective and international resistance for freedom.

The purpose of this study is to understand emancipatory learning by exploring the question, how did the Madres develop the consciousness that helped them move from initial needs-based action to long-term sustained social activism? The notion of social consciousness is also related to this question. Most researchers have ignored this vital learning dimension, which has given a new light to past analysis of Latin American movements: a new form of consciousness fostered by changes in the popular practices of collective action. It is precisely the continually growing consciousness of redefining themselves through their daily practices that I examine in this chapter. The approach to this chapter is partly based on my own learning experience and engagement in the building of resistance movements in Latin America, in which I took an active part during the 1970s and 1980s. Through fifteen years of participation in these movements, I was able to address political concerns in a variety of struggles. I have participated in student action for liberating political prisoners and increasing workers' wages, as well as in national liberation movements.

The Madres movement is particularly significant in terms of the Latin American movement and adult education experience because it challenges us to rethink through the Madres' experiences over the order of participation, conscientization, and social practices that are necessary for the establishment of democracy. The importance of the daily practices of their movement involved "a collective act of creation, a collective signification, a culture" (Escobar, 1992, p. 30). The examination of this focus articulates in the Madres' processes of meaning making through their collective action, undergoing a process of transformation through their everyday actions. Paulo Freire's insights on cultural action and conscientization (1970, 1998) are useful in understanding the learning dimension of the Madres movement. His framework provides the critical tools to reflect on and understand the process that connects learning through praxis and social action, whereby transformative possibilities become a reality. Conscientization occurs through the use of critical tools provided to question and develop a critical standpoint.

The Madres did not know what oppression was when they began their movement. Through the complexities of their everyday actions, however, they learned to understand oppression by voicing resistance against the violence imposed on them and their families. The Madres were thus instinctively participating in and creating a series of mechanisms that needed to be activated for their social, political, and personal transformation. The use of Freire's analytical framework makes sense in terms of understanding this transformation in a critical analysis of their social conscientization. The account of the Madres social movement and consciousness transformation I engage in this chapter is based on the story told by Hebe de Bonafini, founder and president of the movement, who presented the collective recollection of all participating mothers in a seminar held on July 6, 1988, in Argentina. Hebe, like so many Argentinean mothers in the early 1970s, witnessed the disappearance of her two sons and daughter-in-law. Her sacrifices in the name of the movement have been immeasurable. To this day she is leader of the Madres movement, which over the years has built a well-resourced infrastructure, including its own press, library, shelter-home for the poor, a radio program, a literacy coffee house, and the Popular University of the Madres de la Plaza de Mayo. To this day, the Madres continue to meet every Thursday at the Plaza de Mayo.

I will begin this chapter by discussing the Latin American experience from 1970s to 1980s with the upsurge of the democratic social movements. I will then explore Freire's (1998) analytical framework of the levels of conscientization, give examples of the Madres' process of conscientization, and conclude with further reflections on the Madres social movement.

The Madres Movement within the Latin American Experience

There is no doubt that the United States' imperialist actions effected profound changes in Latin America's national and international processes of political and economic development. From the 1930s to the 1970s, U.S. imperialism was challenged by Latin American nationalist, populist, and democratic-socialist regimes and movements (Petras & Veltmeyer, 2001, p. 76): Peron's nationalist movements in Argentina, the Popular Front in Chile, the revolution in Cuba, the radical national revolution in Bolivia, and the national movement for liberation in Nicaragua. The revolutionary changes in governments across Latin America brought about temporary social reforms as well as the institution of varied forms of state capitalism and a state-led development of expanded production (Petras & Veltmeyer).

In Latin America, this epoch, called the "golden age of capitalism" (Marglin & Schor, 1990), was characterized by high rates of economic growth and advancement in economic/social development and was seen as an important "(de)colonialization process." In many of the Latin American countries, the state was converted into the major agency for national development, whose main policy was to implement a new economic model based on "nationalism, industrialization and modernization" (Petras & Veltmeyer, 2001). By the early 1970s, a deep, however, systematic and contradictory, crisis overcame the Latin American structures. The revolutionary regimes began to experience problems arising from the progressive disintegration of their controlled economic societies, which was partly due to large-scale international debts and the rising social and economic gap resulting from the uneven distribution of economic resources and power. High levels of unemployment further contributed to the sociopolitical crisis, causing popular manifestations. It was clear that the status quo had generated forces of opposition and resistance. Across Latin America, revolutionary guerrilla movements promoting national liberation, and peasant and worker resistance movements demanding the right to life, justice, and emancipation proliferated. Political crisis affected Latin America. The issue of who should govern came to the forefront of popular resistance. Many of the governments took on radical authoritarianism to combat these popular and social movements and restructure the state. These radical right-wing regimes were completely unconstitutional. To counteract the mass opposition and resistance across the region, U.S. imperialist policy imposed military regimes in many of the countries, beginning with Peru in 1968, Bolivia in 1971, Chile in 1973, and Argentina in 1976.

The Argentinean Context

In 1976, aided by the mounting economic and political crisis, the military junta headed by General Videla overthrew the Peron government, unleashing large-scale repression (Clark, 2001). The coup's policy was called El Proceso de Reorganizacion National (Process of National Reorganization), designed to establish "order" through unrestrained violence, marking one of the worst periods of human rights violations in modern history. Torture, extrajudicial executions, and disappearances became widespread. If during the 1970s there was an active guerrilla movement in Argentina, by 1979 90 percent of it had been eliminated by the junta's merciless policies (Diario del Juicio, 1985; Navarro, 2001).

But, despite the repression of the guerrillas, social movements grew in Argentina. The idea of national liberation and guerrilla movements as a viable option to overthrow government was replaced by a more democratic view of negotiating political power. These new social movements challenged the state with a different understanding of political practice. Social action grew out of the political crisis and subsequent terror, and actors looked for a political and social space to "produce meanings, negotiate and make decisions" (Escobar & Alvarez, 1992, p. 4) linked to popular interests. In other words, they emerged out of concrete needs (Fals Borda, 1992) with a new understanding of resistance and social change. This new outlook brought forth critical reflection and diverse approaches to the practice of politics.

More than a dozen movements sprouted across Argentina, inspired by social, political, and cultural needs. In the north, the agrarian movement called the Ligas Agrarias demanded land and credit for agricultural needs; in the urban sector, workers' unions had greater political pluralism and connected with other social movements; middle-class women concerned about economic stagnation formed new branches of political parties and organized to protest against inflation. Religious groups mobilized people and resources and marched in the streets in protest of the disappearances and killings. Social movements such as the Association of Argentine Women (AMA) and the Argentina Socialist Confederation (CFA) were created in 1979 demanding their rights. One of its branches, the Socialist Women's Union (UMS), wrote an official manifesto, demanding the restoration of order and stability. The struggle against violation of human rights practices proliferated in Argentina.

At the same time, the Madres' public response to the disappearances of civilians played an important role in preparing the ideological ground for the national and international human rights movements, which joined their cause. The Madres' success was partly due to their constancy. By 1977, the Madres, along with the Grandmothers of the Plaza de Mayo and the Movement of Relatives of the Disappeared, organized mobilizations against the repressive regime. The Women's Multi-sectoral Alliance became the umbrella for these movements. Their mobilized action brought attention to the violations of human rights practices inflicted by the military junta. With the help of other international NGOs and national social movements, the Madres were able to gather some information about the disappeared and report it to the United Nations, who then pressured the Argentinean government to stop the violence. In addition, the French, Italian, and Swedish governments denounced the human rights violations inflicted by the Argentinean government on its citizens. Civil movements in the United States also put pressure on their government to modify its human rights policy. As a result of the unacceptable levels of violence and rising pressure from social movements, in 1978, the U.S. Congress passed a bill eliminating all military assistance to Argentina.

The period between 1976 and 1983 was characterized by indiscriminate violence and intimidation inflicted by the Argentinean military junta across all sectors of society, in Freire's (1998) terms, called a "culture of silence." According to Freire, a culture of silence can only be transformed through social consciousness. I argue that this transformation indeed occurred in Argentina and is particularly evident in the experience of the Madres movement.

Introducing Freire's Framework of Social Action and Conscientization

In his work, Freire underlines the strong relationship that exists between politics and learning. That is, between the ways the state exercises its powers and domination and the constant interplay of civil practices, which include either giving support to state structures or challenging them. This dimension of practice is manifested in social action and conscientization. Conscientization in Freire's words signifies building the awareness to "perceive the social, political and economic contradictions" (1998, p. 35) of the world in which we live. Oppression is the point of reference to conscientization. Recognizing oppression and understanding its causes is an essential precondition to acquiring consciousness. In neo-Marxist theory, oppression is directly linked to the notions of both class and conflict, and when one recognizes the causes of oppression one can act to transform the situation. How does one begin to recognize the causes of oppression?

In Freire's words, "Consciousness is brought about not through an intellectual effort alone, but through praxis—through the authentic union of action and reflection (1998, p. 515). This implies people's participation in the social, political, and economic processes and their commitment to the cause of liberation from oppression. Freire concurs that through participation and commitment one can acquire true democratic values. How did the Madres build democratic politics beyond the personal level? At the beginning, their movement attached little political meaning to finding their disappeared children. So, how did it turn into a collective political struggle within a broader social and political framework? Freire's critical approach helps explain how the Madres' level of consciousness through the years 1976 to 1986 transformed their microactivities and private tragedies into macropolitical concerns and public action.

The Shift Toward a Collective Struggle

Under the military regime in Argentina, and especially between 1976 and 1979, many social movements were eliminated and/or otherwise denied political expression (Cano, 1982). Despite the repressive policies, new voices spoke up, stirring up political manifestations. The Madres was one of the movements that reacted spontaneously to the violent policies of the junta. As Hebe de Bonafini (1988) puts it, their reactions were an expression of sensitive/personal/gut-provoking "interests." In the beginning, they did not have a "class consciousness" against the dictatorship or imperialism; nor did they have awareness, rational or political (Escobar, 1992; Escobar & Alvarez, 1992). Rather, they were driven by personal interest provoked by dramatic events.

Initially, the mothers would individually visit different governmental institutions and political parties to inquire about their children's disappearances. The military identified them as the *locas* (crazy) (Bousquet, 1983, de Bonafini, 1988). They saw the mothers as a nuisance, but not a threat. They scorned at their valor and refused to give concrete answers to their claims. This military/authoritarian tactic encouraged the mothers to seek a new form of expression and a new strategy to attain their goal. Thus, the Madres began to meet collectively and search for ways to challenge the military.

The shift toward a sharpening of political consciousness grew out of their individual confrontation with violence. Their daily searches and individual pleas were left unanswered, but in the process they became aware of the actions of other mothers. Progressively, the Madres began to develop a collective voice. Early evidence of this nascent collective voice is found in a letter of petition signed by sixty mothers asking for their children's whereabouts. In 1976, while at a church, Azucena, one of the unfortunate mothers, said to a group of other mothers: "*Basta!* (This is enough!) Why don't we go to the Plaza with a letter to ask what happened to our children!?" (de Bonafini, 1988). Even though the petition did not receive a response, they experienced collective practices. Further evidence is the Madres' weekly gatherings on Thursdays at the Plaza. Hebe expresses the Madres' collective consciousness as follows:

> We did not know what to do . . . there was no other place we could go to. In the Plaza we felt equal, because we were all the same, because the same thing happened to us, because the enemy was always in the same place. (For these reasons) we resolved to meet in the Plaza. (de Bonafini, 1988)

The Plaza de Mayo thus became their space, which they took because there was nowhere else to go. They instantaneously communicated in the Plaza with other civilians, with whom they shared a common denominator: having a disappeared child. The significance of this action lies in the networking that later led to the mobilization of thousands of women and men against human rights abuses.

The public gatherings in the Plaza created a "soft" organizational structure that was flexible, nonhierarchical, and democratic. Through their passive networking and discussions, they established a flexible decision-making structure, which allowed them to take creative and powerful steps to oppose the military regime and demand their rights. Passive networking implies that the Madres established a network with other mothers without actively or deliberately constructing it. Their nonhierarchical nature allowed them to communicate as equals and to be equally involved in a common identity.

> Why did we feel fine in the Plaza? Well, in other associations we did not feel that we belonged; we did not feel close together (in our cause). There was no office between us. In the Plaza we were equal. We asked each other *"que te pasó?"* (What happened to you?) . . . we were all equal. All of us had their child taken away from us . . . that is why the Plaza consolidated us. (de Bonafini, 1988)

Their weekly gatherings helped them build a line of active communication and cooperation, and their structure shaped their actions against the oppressor. They were seen by the junta as limited and tolerable, while in reality the movement was continuously expanding to unite thousands of Argentinean mothers.

Political Learning During the Early Phase of the Madres Movement

At the beginning, in 1976, the Madres were ignored by other national organizations that had other claims on the government. "We suffered the indifference of the people," said one of the mothers (de Bonafini, 1988). However, their isolation gave the Madres the possibility of moving into particular sites of activity and power. In this sense, "marginalization is not a negative position, but a vector defining access, mobility and possibilities" (Grossberg, 1996). By 1977, they began organizing open meetings in the Plaza. They were a collective body with a clear definition of their goals and without any alliance to political parties. Their shared experiences of violence helped them develop strong ties. Their empowering experience of losing their children enabled them to "see what they did not see before" (Freire, 1998, p. 507). As they shared their experiences they began to question the world. "We were given birth by our children, because after what happened we opened our eyes and started seeing how the world of our children was" (de Bonafini).

This learning was the most significant element of their empowerment. In the process of emerging from silence, the Madres began new forms of actions and struggles. By October of 1977 they began to influence public opinion. Two hundred and thirty-seven mothers demanded a reply from the military junta regarding their disappeared children (Bousquet, 1983). A few days later they handed the military authorities a petition with 24,000 signatures, demanding an investigation into the disappearances, freedom for those who were illegally detained and those detained without trial, and the immediate transfer of those already on trial to civil courts (Feijoo & Gogna, 1990). The Madres who delivered the petition were shot at and approximately 300 were detained for several hours while their documents were recorded. Awareness of the situation gradually expanded to raise the level of recognition of human rights abuse in Argentina.

In December of the same year, Vicenti, one of the founders of the Madres movement, was kidnapped. Many of the Madres were overcome with fear, but despite the repression they continued their struggle (de Bonafini, 1988). Through experiences like these, the Madres were able to become aware of the strength of their collective identity, which gave them power beyond their own knowledge. The Madres confronted the aggressor with unity and cohesion. One example of this occurred when the military seized the Plaza and ordered the Madres to identify themselves. The military obtained their documents and threatened them. After three weeks of intimidation, the Madres decided to hand in 300 identification documents of Madres who were in the Plaza that Thursday. That Thursday, "we were all or none" (de Bonafini).

They reinforced their identity with a white scarf that covered their heads in honor of their disappeared children (de Bonafini). Some of the scarves were painted with the names of their children. In doing this they articulated an alternative discourse (Hernandez, 1997) that bonded them and allowed them to be seen and recognized as a collective. These new forms of struggles are seen as the historical transition to a new consciousness that goes beyond the struggle for personal needs. Further disappearances did not stop the Madres in their struggle. Despite the aggression of the military, they met every Thursday afternoon in the Plaza. Repression was augmented and the Madres Mary Ponce and Esther Balestrino de Cariada were detained, beaten, and sent to prison. The military also brought tear gas and wild dogs to harass the protesting mothers.

> We learned to bring bicarbonate in bottled water and resist the tear gas. We learned more, when being challenged. All this we learned in the Plaza.

> We were women who had never gotten out of our kitchens; learning what so many youth had done before. Struggling to keep that little spot in the Plaza, struggling for that spot of sky that meant the only thing we really had in our lives. (de Bonafini, 1988)

Through their collective experience, they acquired new skills and knowledge. They developed an expertise in finding new ways of resistance and created new forms of contesting the authoritative government. One of their innovative strategies was for all mothers to go voluntarily to prison in the case that one or more were detained by the military.

> It wasn't that they took forty or sixty of us because they wanted. No. We put ourselves to be detained. . . . If one mother goes to prison everyone goes! And if we didn't fit in their truck, we waited for the second one and the third. And if they didn't take us we would go to the police station to present ourselves: "Sir (we would say), I want to be imprisoned with all the mothers!" (de Bonafini, 1988)

The Madres' learning process developed their power to perceive critically. An example of this can be seen when the Madres went to the United States to report to the State Department and legislature about the disappearances. "We had no political preparation and we thought that they (United States) were going to help us" (de Bonafini, 1988). Another example was in 1979 when 150 Madres met with the OAS (Organization of American States) in Buenos Aires.

> We thought that the meeting with the OAS was going to be very important. But absolutely nothing happened. It only served as a target to kill more people, more terror. (de Bonafini, 1988)

The Shift from Defensive Politics to Greater Political Initiative

In 1979, the Madres de la Plaza de Mayo movement became officially constituted as an association. Their governing principles established by the founders stated that they could not join any political party. They believed in internal democracy and practiced open meetings. Their network gradually expanded and linked with other social movements because of the depth of the socioeconomic crisis. The following year they broadened their resistance to join forces with other human rights organizations such as the Grandmothers of the Plaza de Mayo, Relatives of the Disappeared, and People Detained for Political Reasons. They staged their first Resistance March with 150 Madres. The following year they launched a campaign for national mobilization, which aimed at getting the list of detained and disappeared civilians published. Human rights groups and the Grandmothers and Relatives of the Disappeared joined forces with over 5,000 demonstrators (Feijoo &Gogna, 1990).

Over the years they had gained a broader vision and sophistication by participating in national marches, which included the Resistance March, March of the Hands and March of the Scarves, and marches organized by the Permanent Assembly for Human Rights. In 1980, the Madres mobilized and organized their third Resistance March against the government, bringing 15,000 demonstrators. In this march, they announced that "come what may, whoever wins the elections their children must appear alive and those guilty of crimes against the people should be punished" (Feijoo & Gogna, 1990). Gradually their organization shifted from a defensive position to one of greater political initiative. Their protests shifted to proposals bringing forward new arguments that unified different socioeconomic classes and groups. They proposed a strategic initiative in parliament: the creation of a bicameral parliamentary commission in which the Madres and other representatives of human rights organizations would participate when trying and sentencing the assassins of the disappeared.

In 1986, out of desperation, Alfonsin, then civil president of Argentina, sent telegrams to the Madres admitting for the first time in history that the Madres' children were dead. Some of the Madres received coffins with human bones. This was one of the most difficult moments for the Madres. After approximately ten years of struggle, they could not believe that their children were dead. They were surprised and horror-stricken when they received the letters and the coffins. The Madres began to discuss how to respond to the government. There were many contradictions among them. Was it time to stop their struggle and bury their children in peace? Should they accept their children as dead?

> We met, we cried and we were desperate. We took the decision: reject these deaths. Because if we accepted the exhumation of these dead, that they say were caused by confrontation; if we accepted these dead without anyone telling us who sequestered them, without anyone telling us anything, it was like assassinating them again. (de Bonafini, 1988)

Their struggle did not end here. They broadened their original outlook and personal needs. They gathered information about disappearances from the barrios, exchanged information about the disappeared with solidarity groups from other parts of the world, and continued denouncing human rights abuses by the government. Several years later, hundreds of mothers formed the Madres social movement.

> We are going to keep on struggling as our children did. We don't care if they remember the disappeared . . . what interests us is that the social movements accompany us to imitate our children who fought for social justice, for the people and with the people. (de Bonafini, 1988)

Their actions speak louder than words. They undertake dozens of activities each year, including marching and speaking in conferences in Argentina and all over the world. The following account by a Madre speaks clearly of the process of conscientization:

> We have found out that people cannot resolve their cause legally . . . the only way that we have to resolve our cause is through struggle, mobilization and participating actively with the struggle of the people. Governments could make us believe and can tell us that all our problems will be resolved legally, while they attack us with their own made laws. (de Bonafini, 1988)

Conclusion

From opposite ends of the political spectrum, two powerful forces confronted in Latin America: the authoritarian strategy of utilizing violence to eliminate popular mobilization versus social movements and resistance. In analyzing the emergence and development of the Madres resistance, I have presented the multilayered dynamics of their organizational practice: from an individualized struggle to new ways of participation and thinking in their struggle. Through their participation and alternative strategies of resistance they developed the consciousness that helped them move from initial needs-based action to long-term sustained social activism. Using Freire's concepts of action and conscientization, I explored the Madres' contribution to the process of conscientization in today's social struggles.

First, what did the Madres movement demonstrate? It demonstrated the redefinition of the boundaries of the political arena, providing alternative political possibilities. By articulating their concrete goals through protest, mobilization, and political initiative, they provided awareness of the need to transform the state's abusive practices. Their initiatives challenged the state's policies and gave rise to the possibility of building new alternatives for justice. Their practices were a means of recruitment.

Civilians who heard about or participated in the marches were acquainted with the Madres' objectives; their organization became better known and certainly advanced to more popular representation. Their significance is a result of the inner workings of democratic practices, combined with the knowledge of the external conditions of the political and economic situation.

Second, how did the Madres cultivate knowledge of oppression? As the study of the Madres movement demonstrates, their consciousness had an objective and a subjective learning dimension associated with oppression. The objective conditions for cultivating consciousness were generated by understanding the situation of violence and exploitation and the transformation of this understanding into a political project for social action. The subjective conditions are found in their organizational structure, their responses to adverse situations, and solving everyday problems.

The inequalities of oppressed and oppressor is an objective reality of capitalism today. Recognizing them in the everyday experience provides innumerable possibilities to create the conditions for struggle for social justice. Though the objective conditions are necessary, they are not sufficient for cultivating knowledge of oppression. The problem of today is not only acknowledging the objective conditions of oppression but also opposing and critically organizing and directing the process to transform and make better the lives of the oppressed.

The leadership of the social movement must "appeal to the oppressed and exploited and seek to mobilize their discontent into resistance and opposition" (Petras & Veltmeyer, 2001, p. 162). In other words, none of this can be done without democratic participation. The organizational relationship must not be imposed on the people; rather, it must be a participatory relationship with participatory initiatives that will undoubtedly strengthen and contribute to each community member's personal and social development—that of learning to transform consciousness. Not all social movements have the same social/political implications for transforming state's policies. Not all movements have the political support of large mobilizations and protests to make strategic changes to the inequalities in society. But it is important to note that even the most minor forms of resistance provide us with the possibilities to participate and confront the state's unjust policies.

Notes

1. May Square is a central square in Buenos Aires, where the mothers gathered to protest and demand information from the military regime.

2. Intelligence officer in the Argentine navy, known as the Blond Angel of Death. During the rule of Jorge Rafael Videla, Astiz was involved in the deaths of many *desaparecidos.* He was condemned to perpetual life in prison.

References

Amnesty International. (1981). *"Disappearances": A workbook.* New York: Amnesty International USA.

Bousquet, J.P. (1983). *Las Locas de la Plaza de Mayo.* Buenos Aires: El Cid Editor.

Calderon, P., & Reyna, S. (1992). Social movement: Actors, theories, expectations. In A. Escobar & S. Alvarez (Eds.), *The making of social movements in Latin America: Identity, strategy and democracy* (pp. 19–36). Boulder, CO: Westview Press.

Cano, I. (1982). El movimiento feminista argentino en la década del '70. In *Todo es Historia.* Buenos Aires.

Clark, A.M. (2001). *Diplomacy of conscience: Amnesty International and changing human rights norms.* Princeton, NJ: Princeton University Press.

de Bonafini, H. (1988). Conferencia pronunciada el 6 de Julio de 1988, Argentina. Excerpts in the chapter were translated by Baltodano, C.

Diario del Juicio. (1985, December 3). No. 28, 5–8.

Escobar, A. (1992). Culture, economics and politics in Latin American social movement theory and research. In A. Escobar & S. Alvarez (Eds.), *The making of social movements in Latin America: Identity, strategy and democracy* (pp. 62–85). Boulder, CO: Westview Press.

Escobar, A., & Alvarez, S. (1992). Introduction: Theory and protest in Latin America today. In A. Escobar & S. Alvarez (Eds.), *The making of social movements in Latin America: Identity, strategy and democracy* (pp. 1–18). Boulder, CO: Westview Press.

Fals Borda, O. (1992). Social movements and political power in Latin America. In A. Escobar & S. Alvarez (Eds.), *The making of social movements in Latin America: Identity, strategy and democracy* (pp. 303–316). Boulder, CO: Westview Press.

Feijoo, M., & Gogna, M. (1990). Women in the transition to democracy. In E. Jelin (Ed.), J. Zammit, & M. Thomson (Trans.), *Women and social change in Latin America* (pp. 79–114). London Zed Books.

Foley, G. (1999). *Learning in social action: A contribution to understanding informal education.* London: Zed Books.

Freire, P. (1970). *Pedagogy of the oppressed.* New York: Seabury Press.

Freire, P. (1998). Cultural action and conscientization. *Harvard Education Review,* 68 (Winter), 4: Research Library Core.

Grossberg, L. (1996). Identity and cultural studies: Is that all there is? In S. Hall & P. du Gay (Eds.), *Questions of cultural identity* (pp. 87–107). London: Sage.

Guest, I. (1990). *Behind the disappearances: Argentina's dirty war against human rights and the United Nations.* Philadelphia: University of Pennsylvania Press.

Guzmán Bouvard, M. (1994). *Revolutionizing motherhood: The Mothers of the Plaza de Mayo.* Wilmington, DE: SR Books.

Hernandez, A. (1997). *Pedagogy, democracy and feminism: Rethinking the public sphere.* Albany: State University of New York Press.
Jelin, E. (1985). Otros silencios, otras voces: El tiempo de la democratizacion en la Argentina. In G. Calderon (Ed.), *Los movimientos sociales ante la crisis* (pp. 17–44). Universidad de las Naciones Unidas (ONU).
Jelin, E. (1990). Introduction. In E. Jelin (Ed.), J. Zammit & M. Thomson (Trans.), *Women and social change in Latin America* (pp. 1–11). London: Zed Books.
Keck, M., & Sikkink, K. (1998). *Activists beyond borders: Advocacy networks in international politics.* Ithaca, NY Cornell University Press.
Marglin, S., & Schor, J. (1990). *The golden age of capitalism: Reinterpreting the postwar experience.* Oxford: Clarendon Press.
Navarro, M. (2001). The personal is political: The Madres de Plaza de Mayo. In S. Eckstein (Ed.), *Power and popular protest: Latin American social movements* (updated and expanded edition) Berkeley: University of California Press.
Petras, J., & Veltmeyer, H. (2001). *Globalization unmasked: Imperialism in the 21st century.* Halifax, Nova Scotia, Canada: Fernwood.
Sikkink, K. (1993). Human rights, principled issue-networks, and sovereignty in Latin America. *International Organization, 47,* 3.

CHAPTER 15

Adult Education and Development in the Caribbean

Jean Walrond

Introduction

The British Caribbean region, from where my research participants and I emigrated, comprises a group of islands bordering the Caribbean Sea. These are situated east of Central America, north of South America, and south of Florida in the United States of America. For the purpose of this chapter, I am also including Guyana, South America. It is the only English-speaking country in South America, but is culturally, socially, economically, and politically linked to the English-speaking Caribbean region. The region's history includes socioeconomic relationships between a colonizing Eurocentric plantation society and a colonized group consisting of enslaved Africans, indentured South Asians, and the few groups of indigenous Amerindians that are now dispersed throughout the area. The latter occupied the Caribbean area copiously and liberally during the pre-Columbian epoch.

In the context of the history of the black[1,2] Caribbean region, education and development are two nuanced and complex concepts. Beckford (1976) contends that "during slavery educational opportunities were restricted" (p. 37). After emancipation, the enslavers/colonizers provided the slaves improved educational opportunities but still geared them toward skills useful to the plantations. Nyerere (1968) states that the purpose of education, after slavery and during colonialism, "was motivated by a desire to inculcate the values of the colonial society and to train individuals for the service of the colonial state" (p. 269). Blacks rejected this notion, viewing education as enabling social mobility and a constituent of European lifestyle (Miller, 1976). As such, education was supposed to differentiate those individuals "who had 'made it' from those who had not" within this highly socially stratified society (Beckford, p. 38). Similarly, the purpose of adult education for people of the Caribbean is challenging to define, but its meaning can be aligned to Friboulet's (2005) description. He provides an inclusive definition of education, as a learning process that enables societal, interpersonal, and psychological empowerment; participation in development; and lifelong learning.

Development is associated with economics, growth, blossoming progress, and expansion (Rist, 1999). But in terms of emancipation, the emotional associative mapped to development were negative psychological reminders of slavery. These were the eradication of native African languages, culture, and ways of knowing, all important for individual and community development. Julius Nyerere's (1968) definition of development would have been quite fitting to describe the freed slaves' perspective of the term. He defines development as the state of being where one has acquired the values of colonial society and is able to be of service to the colonial state. From this axiological position, how should those who live with a legacy of enslavement value education as prescribed by the heretofore enslavers? As well, how would they establish their own goals for education? Nyerere also writes that "school" or formal education is not a necessary organization for development. He states: "[The] purpose [of education] is to transmit from one generation to the next the accumulated wisdom and knowledge of the society, and to prepare the young people for their future membership of the society and their active participation in its maintenance or development" (p. 268). On this basis, one can understand the dilemma that faced blacks in the Caribbean in terms of development and education. In this chapter, informal, nonformal, and formal forms of education are included under the rubric of basic adult education, and an analysis will demonstrate that these were critical to the development of a Caribbean social, cultural, and political sensibility.

Education for the People: Formal Adult Education and Development, Fostering the Agricultural Sector

Formal adult education for blacks in the Caribbean was introduced to facilitate the needs of the plantations. Thus, the emphasis was on developing and implementing agricultural programs and other trades that were useful to the plantations. Bacchus (2005) states that these schools were established at Botanical Gardens and the program offered courses in the knowledge of identifying and naming tropical plant species, plant and animal husbandry, disease control, propagation of plants, farm management, and the maintenance of a nursery. Often students became apprentices and some even qualified for bursaries. Many others participated in programs that were instituted at government or privately owned farms. Once the period of apprenticeship was over, graduates found employment in sugarcane plantations, orange groves, cocoa estates, coffee plantations, and coconut plantations. On islands such as St. Vincent, which specialized in the cultivation of specific products such as Sea Island cotton and arrowroot, the apprentice programs were tailored to the care and production of these crops. Ownership of the farms usually remained in the hands of the white farmers or foreign companies, and the highest position to which blacks could aspire were overseers who usually lived on the farms, estates, or plantations; received free housing; and were responsible for the entire operation of the estate.

Agriculture continued to be the focal point of development in the Caribbean. This was reflected in the importance placed on providing education in this area. Bacchus (2005) writes:

> From the early 20th century, Trinidad had two agricultural training centres that provided students with a sound practical education by making them take an active part in ordinary estate work. In addition, a correspondence home-reading course was offered by the Agricultural Department for estate employees and, on successful completion, students were awarded a certificate of achievement. (p. 258)

In 1946–1947 the Imperial College of Tropical Agriculture, at St. Augustine, Trinidad and Tobago, was inaugurated with the responsibility for both teaching and research in agriculture. In 1960, the college was expanded to become the University of the West Indies, St. Augustine campus. With degree-granting status in agriculture, science, and engineering disciplines, this campus provided tertiary-level education to the entire Caribbean region.

Education Toward the Development of Other Economic Sectors

As we shall see, agriculture was just one sector of the economy that contributed to the development of the region. Other sectors were in the area of arts and crafts[3] and the service industries. But for the most part, there continued to be a shortage of skilled workers for these other sectors due to lack of funding for the necessary educational programs. Only the larger British Caribbean countries such as Jamaica and Trinidad and Tobago had limited success, but they were still not able to meet the demands for skilled workers. Bacchus (2005) writes: "The cost of establishing and operating technical schools was too high for most colonies, though Jamaica set up the Kingston Technical College in 1911, the Jamaica Institute of Technology in 1958, and Trinidad [and Tobago] had the Victoria Institute" (p. 280). Trinidad and Tobago and other smaller countries also established Boards of Industrial Training, which were responsible for training, examining, and certifying local crafts workers (Bacchus). These boards were responsible for providing night classes in pattern making, furriery, building construction, bookbinding, shoemaking, workshop arithmetic, boatbuilding, and other crafts essential to the development of the region. In some cases, students who demonstrated excellent competence in their craft were selected for apprenticeship in the master craftsmen category, thus qualifying for higher wages and benefits and to become supervisors. Females who had completed six years of schooling qualified to study courses in commercial subjects. These were also under the auspices of the Board. Sometimes these classes were cancelled when qualified teachers were not available to teach the courses.

Although the West Indian Conference of 1921 recommended that regional technical schools be established throughout the Caribbean islands to provide part-time classes in arts and crafts, these recommendations were never fully attained because of lack of adequate financial resources. As a result of these challenges, the shortage of trained technical and professional staff persisted, and some employers instituted their own apprenticeship training programs to provide employees for very crucial services. These programs were a combination of on-the-job training and night classes. Pharmacy, nursing, and other medical technicians were in this category. Training and certification were controlled by the Medical Board by an Act of parliament. My father and many of his colleagues qualified as pharmacists under this system in the 1930s.

The Jamaica institute of Technology was renamed the College of Arts, Science and Technology (CAST) in 1959 with continued expansion to its student body and programs. In 1964 CAST's incorporation was recognized by an Act of Parliament, and in 1985 it became a degree-granting institution. Today, with pedagogy modelled on the British polytechnic system, it boasts a student population of more than 6,000 and offers at least 100 courses at certificate, diploma, and degree levels (University of Technology, Jamaica, 2003). In Trinidad and Tobago the

John S. Donaldson Technical Institute, inaugurated in 1961 by Dr. Eric Williams, was "seen as the means of fulfilling the manpower needs of the independent nation of [Trinidad and Tobago]" (p. 2). This institute's platform states that its

> preliminary objectives included self sufficiency, economic growth, alleviation of the unemployment problem through the application of Technology with accompanying skills, reduction of the dependence on other countries and generally improving the living conditions of the people of Trinidad and Tobago. (University of the West Indies Tertiary Level Institutions Unit, 1998, p. 2)

These are some of the ways the people in the Caribbean obtained formal education once they had successfully completed their school leaving certificate exam at about ages fifteen to seventeen (Miller, 1976). The preceding programs of formal adult education provided individual development opportunities to a few people in the Caribbean. Many more inhabitants were struggling under the yoke and vicious cycle of poverty and lack of education, a situation that would be satiated with political dissatisfaction of colonialism.

Informal and Nonformal Education: Contextual Definition

Barakett and Cleghorn (2000) define informal education as everyday education that normally starts at birth and is most likely to occur in informal settings such as the home. In contrast, they state that nonformal education "refers to organized instruction that takes place outside of the school settings (e.g., [boy scouts,] girl scouts, music lessons, sports groups)" (p. 2). Within the Caribbean these two modes of adult education are difficult to separate, as settings that are constructed for informal reasons are invariably transformed to opportunities for collective nonformal teaching and learning. This was particularly true in the Caribbean during periods of nationalistic transitions and debates.

Generally, debates about colonialism, anticolonialism, and postcolonialism involve deconstructing a literary or historic text. I was born and raised in the Caribbean during these transition periods and viewed this discourse from a critical theorist's axiological position. I often struggle with viewing colonialism, anticolonialism, and postcolonialism in terms of discourses that must be deconstructed. For me, my experiences with these notions are lived experiences rooted in informal education and public pedagogy. Having said this, I understand that these experiences and context ground my perspectives on the notion that "propositional, transactional knowing is instrumentally valuable, as a means to social emancipation which is an end in itself [and] is intrinsically valuable" (Guba & Lincoln, 2005, p. 198). Similarly, my data show that my research participants also recognized that they were schooled during the colonialism era as much of their discussion centered on the activities that were done to enhance and promote consciousness. In some instances, the participants actually spoke about political awareness while one participant in particular gave a very graphic account of the impact political awareness and action had on her education.

The Media and Its Role in Adult Education Pedagogy

In my family we all had the habit of listening to preindependence debates as these were broadcast on the radio every night when parliament was in session. Not only did colonialism produce vituperative debates but these were also opportunities for witnessing the historical processes of political independence as it birthed before our very eyes. Also, this culture was possible because of the public pedagogical work that nationalist-bent political parties undertook in an effort to educate the citizens of Caribbean. In my case, one such party was the People's National Movement (PNM), the first national political party in Trinidad and Tobago. The party educated the citizenry through these parliamentary debates and meetings in town squares throughout the country. In Eric Williams's[4] own words,

> The People's National Movement made the first plank in its platform the political education of the people. It organised what has now become famous in many parts of the world, the University of Woodford Square, with constituent colleges in most of the principal centres of population in the country. The political education dispensed to the population in these centres of political learning was of a high order and concentrated from the outset on placing Trinidad and Tobago within the current of the great international movements for democracy and self-government. The electorate of the country was able to see and understand its problems in the context of the ancient Athenian democracy or the federal systems of the United States and Switzerland, in the context of the great anti-colonial movements of Nehru and Nkruma, and in the context of the long and depressing history of colonialism in Trinidad and Tobago and the West Indies. The voter in Trinidad and Tobago was quite suddenly invested with a dignity for which he was obviously, by his response, well-fitted. (1962/1993, p. 244)

This was one role undertaken by social movements in the Caribbean, and my research participants and I were recipients of these experiences while growing up in the 1950s, 1960s, and 1970s. These were arenas for contested debates as anyone could have the opportunity to express their opinion, and crowds gathered there every night to listen to the cabinet debates from the Red House and to articulate or, better, to argue their positions in postanalysis.

Public Speech and Its Role in Adult Education Pedagogy

My other recollections of adult education discourse were incidents I observed as a child on my way to school. I grew up in the oilfield town of Point Fortin in Trinidad and Tobago. On my way to school, I frequently heard Tuber Uriah Buzz Butler, a union leader and politician, pontificating about the unfair labor practices that existed at Trinidad and Tobago's Shell and Texaco oil companies. I concur with Williams's (1962/1993) description of Butler as the politician who "brought the inarticulate masses on a national scale on to the political stage" (p. 243), and assert that this assessment of anticolonial sentiment was rampant throughout the Caribbean. Williams's own approach to the anticolonial discourse was altogether different.

As leader of the PNM, Williams was not only a learned historian but also a skilled verbalist, and his use of the phrase "Massa Day Done" (1962/1993, p. 240) provided him an opportunity to educate and instruct the masses to interrupt the prevailing discourse of servitude. His words rallied the masses to apply a new and opposite perspective to their relationship with the colonial masters. He did not recant the phrase or apologize for saying it to the press. Instead, he repeated it more than 95 times in a subsequent speech, which was in response to the media's call for an apology. Those who, perhaps, would have missed the printed comment or could not read it now had it resonating in their psyche. In this way, Williams used the spoken discourse successfully to stimulate critical consciousness and to effect collective emancipation. Freire (1970/2005) writes: "Populist manifestations perhaps best exemplify this type of behavior by the oppressed, who, by identifying with charismatic leaders, come to feel that they themselves are active and effective" (p. 78) in their emancipation. Williams **used** hyperbole and exaggeration effectively in his statements to make his case. For example, he did not provide any evidence or warrant that all "Massa's children were educated at the best schools and at the best Universities" (Williams, 1962/1993, p. 240). He did not provide evidence that they went to

Oxford and Cambridge, but to many in his audience it was a plausible statement. After all, masters' children were not being educated at the poorly equipped and poorly managed schools in the Caribbean. As Bakhtin (1965/1984) attests, this type of speech genre is useful and common in the marketplace where "the enormous accumulation of superlatives, typical of marketplace advertising [is] the characteristic method of testifying to the speaker's honesty" (p. 162). As well, Williams, in referring to Woodford Square as the University of Woodford Square, converted one of the bastions of colonial rule to the seat of learning for the layman and laywoman. By addressing the people in a public square, he demonstrated that he was one with the people because "in the town square, a special form of free and familiar contact . . . [reigns] among people who are usually divided by the barriers of caste, property, profession and age" (p. 12).

Referencing Foucault, who writes that a space can be an "important sign and instrument of autonomy" (1993, p. 136), I argue that Williams used the space, Woodford Square, as a symbol of power and knowledge for the people of Trinidad and Tobago and the West Indies. I cite his use of the phrase "the architect of the University of Woodford Square" to support my contention. Also, just as Foucault contends that it is "somewhat arbitrary to dislocate the effective practice of freedom by people, the practice of social relations, and the spatial distribution in which they find themselves" (p. 136), Williams and the PNM party understood the educative opportunities that were present when over 20,000 would gather there to listen to the speeches, and they used those occasions to their advantage (Cudjoe, 1993).

Williams mentions that the platform of the PNM party places free secondary education for all high on its agenda. This will "safeguard equality of opportunity and a career open to talent" (Cudjoe, 1993, p. 254). The use of the pronouns "our" and "we" implies that he is taking personal responsibility for providing free secondary education and, likewise, for the social and economic welfare of those who apply themselves scholastically.

Williams uses the final words of his "Massa Day Done" speech to encourage and rally the people of Trinidad and Tobago toward independence. Cudjoe (1993) writes:

> Although "Massa Day Done" signifies the acme of Williams's political career as an orator and the embodiment of a people's hopes and desires, it also signified the culmination of a particular manner of rendering the society's experiences and returning a people to its own voice. (pp. 82–83)

There is no doubt that the PNM's intentions, as stated in its manifesto, were the education of the people. This clarification can be found in some of Williams's other speeches. For example, at the party's third annual convention, he spoke about the efforts that the party had made toward the political education of Caribbean people. In speaking of his party's accomplishments, or "miracles" as he termed them, he said:

> The third "miracle" is the establishment of the Party Forum, the University of Woodford Square. Through this forum it has brought political education to the people of the West Indies and introduced a new technique of cold intellectual political analysis based on reasoning and facts as against the empty emotionalism of the past. . . . The international recognition of the University of Woodford Square is illustrated by the front page photograph in the current issue of P. N. M. Weekly, which reproduces a German picture with supporting text of "Universitat von Woodford-Square." (Cudjoe, 1993, p. 208)

In his 1956 speech, titled "The Chaguaramas Declaration: Perspectives for the New Society," where in the party's manifesto was laid out, Williams emphasized the importance of education for the child by stating thus:

> The personality of the child can also be shaped in the school. Our educational system must seek to shape a creative, innovative being who is prepared to evolve solutions to problems for himself. (Cudjoe, 1993, p. 303)

And again, to relate the value of education to the development of a new nation, he was inclusive in his ideas as to the type of education required. He stated the following to emphasize this point:

> Social attitudes in the society at large should be influences in order to achieve, at all levels, the acceptance of the products of all types of education—technical, vocational and academic. (Cudjoe, 1993, p. 303)

The PNM party not only focused on nonformal educational initiatives but also sought to address curricula matters in an effort to make it more relevant to the students in the country. The People's Charter, as their organizational manifesto was titled, attacked "the uncritical imposition of alien standards and curricula unrelated to local needs developed in a different climate for people with a different history and different traditions" (Sutton, 1981, p. 14). It continued:

> The educational system needed today is, by contrast, one which is designed:
>
> to satisfy the legitimate demand of the people for education as their democratic right
>
> to relate education to the local environment of local needs
>
> to produce the highly trained workers and the responsible citizens needed in the age of self-government
>
> to ease the strain on the labour market by keeping juvenile workers out of it and retraining them in school. (p. 14)

Supporting these initiatives were an increase in free exhibition places that increased the opportunities for students to attend school. Thus, even though these initiatives were too late for some of my research participants, others who followed were able to take advantage of them. These were some of the speeches that encouraged students and families to emphasize education, and they also informed them that the PNM's responsibility would be education for all.

Youth Camps as Sites for Adult Education Pedagogy

Informal education was also used to correct the low levels of education among male youth who had dropped out of school without having their school leaving certification, which was the minimal credential for obtaining meaningful employment. Public pedagogy was conducted in community centers, town and village squares, and, as Lance (pseudonym) mentioned, in boys' campsites. This form of emancipatory work ensured the PNM's success in politics, but it was also instrumental in helping my research participants to value education. Although Lance indicated in our discussion that he was dismissed from the camp for disobedience, he admitted to me that he now values the attempts the PNM youth workers made to educate him at that time. Such leaders in the Caribbean encouraged my research participants. Errol's (pseudonym) two mentors were Ralph Gonzales (his schoolteacher) and the present prime minister of St. Vincent and Guyanese scholar and political activist Walter Rodney, whose books include *How Europe Underdeveloped Africa* and *Groundings of My Brothers.*

The public pedagogy, through which my research participants were informed, inferred that they were to be civil-minded citizens and that their leaders were going to support them in their efforts to succeed. Williams was an excellent role model to purport those ideas because he was a Trinidad Island scholar and an Oxford scholar who graduated first in his class. By including the lines from Lincoln's Declaration of Independence in his speech, Williams attempted to inspire everyone, including young people, to take up the fight for personal emancipation. Examples of the altruistic phrases he used in his passionate discourses were "with malice toward none" and "with charity for all," "with firmness in the right, as God gives us to see the right," and "let us strive on to finish the work we are in."

Another era in Caribbean politics to which one of my research participants referred as being influential in her education was that of Maurice Bishop and the New Jewel Movement in Grenada. What follows is an excerpt of Bishop's July 2, 1979, oration on education:

> Perhaps the worst crime that colonialism left our country, has indeed left all former colonies is the education system. This was so because the way in which that system developed, the way in which that system was used, was to teach our people an attitude of self-hate, to get us to abandon our history, our culture, our values. To get us to accept the principles of white superiority, to destroy our confidence, to stifle our creativity, to perpetuate in our society class privilege and class difference.
>
> The colonial masters recognised very early on that if you get a subject people to think like they do, to forget their own history and their own culture, to develop a system of education that is going to have relevance to our outward needs and be almost entirely irrelevant to our internal needs, then they have already won the job of keeping us in perpetual domination and exploitation. Our educational process, therefore, was used mainly as a tool of the ruling elite (Bishop, 1982, p. 79)

In 1979, Maurice Bishop assumed power after the coup d'etat that ousted Sir Eric Gairy, the charismatic, long-standing, but corrupt leader of Grenada. Parry, Sherlock, and Maingot (1987) write:

> While the movement against the increasingly eccentric and incompetent Gairy had wide support it was actually led by a small group of middle class radicals called the New Jewel Movement. Their leader was a charismatic young lawyer, Maurice Bishop, whose Marxist ideology was not revealed to the Grenadians until after the coup d'etat, but whose wide appeal and popularity was undisputed. (p. 289)

Bishop's office as prime minister was quite tumultuous and revolutionary. He was eventually overthrown by another **coup d'etat** and executed in 1983 by Bernard Cord, and the events subsequently led to the invasion of Grenada by the U.S. military (Bishop, 1982; Parry et al.,

1987). Maurice Bishop's speech, on July 2, 1979, at the start of the National Education Conference, is worth analysis to locate the key themes in order "to draw a picture of the pre-suppositions and meanings that constitute the cultural world of which the textual material is a specimen" (Peräkylä, 2005, p. 870). The speech provides a historical review of education for the enslaved. As Bishop indicated, this was nonexistent during slavery because the slave owners felt "slaves were not required to know how to read or write, nor were they required to think for themselves" (Bishop, 1982, p. 80). He continues that when slavery was abolished in 1834, the Negro Education Grant was established, but this "only ensured the continued exploitation of the Grenadian" (p. 80). As Bishop says, some 140 years later very little improvement is apparent: there is very high illiteracy, especially in the rural areas; the country continues to be producers of raw material for their former colonial masters; 30 percent of elementary teachers are trained, and even worse, only 7 percent of secondary teachers are trained. In consideration of these facts, he continues:

> It is easy for any government, it certainly will be easy for the People's Revolutionary Government, to proclaim the principle of free education for all. And this we are of course very happy to do. But it is one thing to say free education, it is another thing to say how are we going to pay for that free education. Where is the money going to come from? Where are the resources going to come from that we are certainly going to need to run schools, train teachers better, provide a more relevant form of education, and all free of cost? (Bishop, 1982, p. 85)

This passage reveals someone who is passionately seeking answers as to how to educate the people. Bishop recognizes that literacy is a problem, and he not only seeks to educate the youth but also strives to make the whole nation literate. To achieve this, Bishop appeals to the people to volunteer to eradicate illiteracy. He continues:

> All of us are going to have to strive to become teachers on the job and off the job. All of us are going to have to try to get down to the important task of raising the literacy standard, providing all our people with the basic opportunity of being at least able to read and to write. (p. 85)

He ends his speech with an even stronger appeal to his comrades by addressing them as his brothers and sisters and reiterating the following:

> So to summarise, Sisters and Brothers, we must move to wipe out illiteracy, we must move to develop a system of work and study in the schools,

> we must move to make all of who are capable of being such teachers, develop the concepts of taking education into the countryside on a voluntary basis to those of our unfortunate sisters and brothers who are not even able to come to the town to get that education.
>
> We must use the educational system and process as a means of preparing the new man for the new life in the new society we are trying to build. (p. 86)

Clearly, these excerpts provide ample understanding of the commitment that some leaders in the Caribbean have to the education of the masses. These passages show that by first explaining the situation and by providing the people with empowering opportunities to become teachers, Bishop recognizes that the people can participate in their own emancipation. He also recognizes that as a nation they must see themselves as agents of their own success as they cannot rely on outside influences to provide any assistance in the education of his people.

Much of Bishop's grassroots approach to eliminating adult illiteracy was modeled after Paulo Freire's framework. In fact, Freire travelled at least twice to Grenada, in December 1979 and again in February 1980, to work with the new government of Grenada and their Ministry of Education to implement the literacy program (Freire, 1998).

Informal Education

Evidence of our colonial education came up constantly during the conversations. In terms of formal curricula, this is what Ursula (pseudonym) shared: *"I probably knew more about Canada and England than I did about the Caribbean. . . . Every thing was sort of colonial Very, very colonial." (personal communication, 2004)*

Other evidences of our colonial education are expressed in the poetry we were forced to memorize and consequently remember today some forty years after the fact. Postcolonial theorists such as C.L.R. James (1963/1983), JanMohamed (1995), Lamming (1995), and Said (1979), who emphasize the importance of the literary text "as a site of cultural control and as a highly effective instrument for the determination of the 'native' by fixing him/her under the sign of the other" (Ashcroft, Griffiths, & Tiffin, 1995, pp. 8–9), confirm this belief.

JanMohamed (1995) provides a theoretical analysis when he states: "Colonialist literature is an exploration and a representation of a world at the boundaries of civilization that has not (yet) been domesticated by European signification or codified in detail by its ideology" (p. 18).

In my dissertation, I used my participants' voices to add other layers of understanding to the ways Caribbean people constructed knowledge from the colonial texts they studied (Walrond, 2007). Further, my analysis concluded that people in the Caribbean contested education by using their understanding of the world to contextualize and critique the prevailing discourse (Walrond). Informal public pedagogy, in the form of the musical genre—the calypso—provided these other levels of understanding of the dominant discourse and these were accessible to everyone. Calypsos were lyrical, catchy rhythms with double entendre expressions or flagrant criticisms. Critiques of the colonial education system were expressed in our calypsos. For example, Rosemary (pseudonym) expressed the following statement about her resistance at having to learn literature, which was irrelevant to her reality.

> *Poetry, we had to learn [it]*
> *And stand in front of the school, [or] class*
> *And say [it]*
> *I rebelled*
> *I could write poetry now if I wanted to*
> *But to learn it, forget it.* (personal conversation, 2004)

Popular culture genres provided counternarratives to the education discourse, and these likely caused us to critique and question the literature we studied in school. For example, a portion of the popular calypso *"Dan is the man in the van,"* written and sung by Francisco Slinger (1963), illustrates this point. He sings:

> *According to the education you get when you small*
> *You . . . [could] grow up with true ambition and respect from one and all!*
> *But in my days in school they teach me like a fool*
> *The things they teach me I should be a block-headed mule.* (p. 1)

Herein the calypso singer makes a statement about his education, which is followed by a string of titles to the poetry, nonsense rhymes, and stories he endured as a schoolboy. The cultural pedagogy that is exemplified in this verse is the nuanced way the calypso is employed to educate even when it functions "to preserve the fervour of the [Carnival] festival, and the life-force without which the masquerade itself, and even the more serious music that resonates at its periphery or beneath its mask of gaiety, would cease to exist" (Rohlehr, 1998, pp. 86–87).

Conclusion

This chapter analyzes three types of education that were prevalent in the Caribbean. Formal adult education improved the conditions of those who were able to access it, and in so doing it contributed to the development of the Caribbean. Blacks in the Caribbean did not wish to get an education to help develop the Caribbean; instead they viewed it as a vehicle for their own upward social mobility. Some of our leaders who were educated under the colonial system understood that education was inextricably linked to nation building in a postcolonial world. They instilled critical consciousness into the hearts and minds of the masses with public pedagogy in the town squares. We have to be mindful that "critical consciousness may lead to disorder" (Freire, 1970/2005, p. 35). This was indeed the case with Grenada. Finally, the history of enslaved Africans is routed in a discourse where double entendre and pun in our musical genre helped us to "come to voice" when under the master's gaze. This speech was perfected and is still used today to critique structural systems. These three types of educational discourse had contributed to the economic, social, cultural, and political aspects of people of Caribbean heritage who should bring this knowledge to educational discourses in Alberta's education system.

Notes

1. The term *black* is used to describe phenotypical qualities that usually pertain to people of African decent.
2. The research participants for this chapter self-identified as blacks of Caribbean heritage.
3. Crafts included such skills as carpentry, masonry, and plumbing.
4. Dr. Eric Williams became the first black prime minister of Trinidad and Tobago, which he led to political independence in 1962.

References

Ashcroft, B., Griffiths, G., & Tiffin, H. (1995). Introduction. In B. Ashcroft, G. Griffiths, & H. Tiffin (Eds.), *The post-colonial studies reader* (pp. 7–11). New York: Routledge.

Bacchus, M.K. (2005). *Education for economic, social and political development in the British Caribbean colonies from 1896 to 1945*. London, ON: The Althouse Press.

Bakhtin, M. (1984). *Rabelais and his world* (H. Iswolsky, Trans.). Bloomington, IN: Indianas University Press. (Original work published 1965)

Barakett, J., & Cleghorn, A. (2000). *Sociology of education: An introductory view from Canada.* Scarborough, ON, Canada: Prentice Hall, Allyn and Bacon.
Beckford, G.L. (1976). Plantation society: Towards a general theory of Caribbean society. In P.M.E. Figueroa & G. Persaud (Eds.), *Sociology of education: A Caribbean reader* (pp. 30–46). Toronto, ON: Oxford University Press.
Bishop, M. (1982). *Forward ever! Three years of Grenadian revolution: Speeches of Maurice Bishop.* Sydney, Australia: Pathfinder Press.
Cudjoe, S.R. (Ed.). (1993). *Eric E. Williams speaks: Essays on colonialism and independence.* Wellesley, MA: Calaloux Publications.
Foucault, M. (1993). Space, power and knowledge. In S. During (Ed.), *The cultural studies reader* (2nd ed., pp. 134–141). New York: Routledge.
Freire, P. (1998). *Pedagogy of hope: Reliving pedagogy of the oppressed.* New York: Continuum.
Freire, P. (2005). *Pedagogy of the oppressed* (30th anniversary ed., M.B. Ramos, Trans.). New York: Continuum. (Original work published 1970)
Friboulet, J.J. (2005). Measuring a cultural right: The effectiveness of the right to education. In A. Osman (Ed.), *Achieving education for all: The case for non-formal education. Report of a symposium on the implementation of alternative approaches in the context of quality education for all* (pp. 17–28). London, UK: Commonwealth Secretariat.
Guba, E.G., & Lincoln, Y.S. (2005). Paradigmatic controversies, contradictions, and emerging confluences. In N.K. Denzin & Y.S. Lincoln (Eds.), *Handbook of qualitative research* (3rd ed., pp. 191–215). Thousand Oaks, CA: Sage.
James, C.L.R. (1983). *Beyond a boundary.* Durham, NC: Duke University Press. (Original work published 1963)
JanMohamed, A.R. (1995). The economy of Manichean allegory. In B. Ashcroft, G. Griffiths, & H. Tiffin (Eds.), *The post-colonial studies reader* (pp. 18–23). New York: Routledge.
Lamming, G. (1995). The occasion for speaking. In B. Ashcroft, G. Griffiths, & H. Tiffin (Eds.), *The post-colonial studies reader* (pp. 12–17). New York: Routledge.
Miller, E. (1976). Education and society in Jamaica. In P.M.E. Figueroa & G. Persaud (Eds.), *Sociology of education: A Caribbean reader* (pp. 47–66). New York: Oxford University Press.
Nyerere, J. (1968). Education for self-reliance. In J. Nyerere (Ed.), *Freedom and socialism: A selection from writings and speeches, 1963–67* (pp. 267–290). London: Oxford University Press.
Parry, J.H., Sherlock, P., & Maingot, A. (1987). *A short history of the West Indies* (4th ed.). London: Macmillan.
Peräkylä, A. (2005). Analyzing talk and text. In N.K. Denzin & Y.S. Lincoln (Eds.), *Handbook of qualitative research* (3rd ed., pp. 869–881). Thousand Oaks, CA: Sage.
Rist, G. (1999). *The history of development: From western origins to global faith.* New York: Zed Books.
Rohlehr, G. (1998). "We getting the kaiso that we deserve": Calypso and the world music market. *The Drama Review, 42*(3) (T159), 82–95.

Said, E. (1979). *Orientalism.* New York: Vintage Books.
Slinger, F. Mighty Sparrow. (1963). Dan is the man in the van, DF66; in Calypso Sparrow KW, 120.
Sutton, P.K. (1981). *Forged from the love of liberty: Selected speeches of Dr. Eric Williams.* Trinidad, West Indies: Longman Caribbean.
University of Technology, Jamaica. (2003) *About UTech.* Retrieved December 27, 2006, from http://www.utechjamaica.edu.jm/About/history.html
University of the West Indies Tertiary Level Institutions Unit. (1998). *John S. Donaldson Technical Institute TLIU, April 1998.* Retrieved December 27, 2006, from http://www.cavehill.uwi.edu/tliu/cc/jdti.htm
Walrond, J. (2007). *Caribbean Canadian culture matters: Our perspectives of education "back home" and in Edmonton's schools.* Unpublished doctoral dissertation, University of Alberta, Edmonton.
Williams, E.E. (1993). Massa day done. In S.R. Cudjoe (Ed.), *Eric E. Williams speaks: Essays on colonialism and independence* (pp. 237–264). Wellesley, MA: Calaloux Publications. (Original work published 1962)

Index